Sebastian Schulz

Phase-field simulations of multi-component solidification and coarsening based on thermodynamic datasets

Schriftenreihe
des Instituts für Angewandte Materialien
Band 65

Karlsruher Institut für Technologie (KIT)
Institut für Angewandte Materialien (IAM)

Eine Übersicht aller bisher in dieser Schriftenreihe erschienenen Bände finden Sie am Ende des Buches.

Phase-field simulations of multi-component solidification and coarsening based on thermodynamic datasets

by
Sebastian Schulz

Scientific Publishing

Dissertation, Karlsruher Institut für Technologie (KIT)
Fakultät für Maschinenbau
Tag der mündlichen Prüfung: 13. Mai 2016

Impressum

KIT Scientific Publishing

Karlsruher Institut für Technologie (KIT)
KIT Scientific Publishing
Straße am Forum 2
D-76131 Karlsruhe

KIT Scientific Publishing is a registered trademark of Karlsruhe Institute of Technology. Reprint using the book cover is not allowed.

www.ksp.kit.edu

Print on Demand 2017 – Gedruckt auf FSC-zertifiziertem Papier

ISSN 2192-9963
ISBN 978-3-7315-0618-8
DOI 10.5445/KSP/1000063920

Phase-field simulations of multi-component solidification and coarsening based on thermodynamic datasets

Zur Erlangung des akademischen Grades
Doktor der Ingenieurwissenschaften
der Fakultät für Maschinenbau
Karlsruher Institut für Technologie (KIT)

genehmigte
Dissertation
von

Dipl.-Ing. (FH) Sebastian Schulz, M.Sc.

Tag der mündlichen Prüfung:	13. Mai 2016
Hauptreferent:	Prof. Dr. rer. nat. Britta Nestler
Korreferent:	Prof. Dr. Hans Jürgen Seifert

Kurzfassung

Die Nutzung von thermodynamischen und kinetischen Daten spielt eine zentrale Rolle im Bereich der quantitativen Phasenfeld-Modellierung. Die vorliegende Dissertation befasst sich mit diesem Thema auf theoretischer Ebene und beschreibt außerdem die praktische Anwendung solcher Daten für unterschiedliche Phasenfeld-Studien. Im ersten Teil der Dissertation werden die Grundlagen der Thermodynamik und Diffusion behandelt, wobei der Schwerpunkt auf Modellen zur Beschreibung der freien Enthalpien liegt. Diese physikalischen Größen sind grundlegend in der rechnergestützten Thermodynamik, welche zusammen mit der Methode der rechnergestützten Kinetik beschrieben wird. Ein weiteres Kapitel ist dem quantitativen Phasenfeld-Modell basierend auf großkanonischen Potentialen gewidmet, welches für die präsentierten Phasenfeld-Studien Verwendung findet. Da solche Simulationen lediglich einen kleinen Konzentrations- und Temperatur-Bereich abdecken, können sie bezüglich des Rechenaufwands durch die Verwendung von vereinfachten Ausdrücken für die thermodynamischen Funktionen optimiert werden. Unterschiedliche Strategien zur Modellierung dieser Funktionen werden beschrieben und für Material-Systeme mit besonderen Eigenheiten, wie zum Beispiel stöchiometrischen oder pseudo-binären Phasen, angewendet. Der Einsatz von Taylor-Entwicklungen und der Methode der kleinsten Quadrate wird hinsichtlich der Abweichungen von den ursprünglichen Formulierungen diskutiert.

Diese Dissertation enthält eine Stabilitäts-Analyse einphasiger Wachstumsfronten von Aluminium-Silizium-Legierungen in Abhängigkeit unterschiedlicher Material- und Prozessparameter. Für ein System mit zwei Komponenten und isotropen Oberflächen-Energien wird die Stabilität von planaren Fronten untersucht. Die simulierten Wachstumsraten sinusoidaler Störungen stimmen gut mit der Mullins-Sekerka-Theorie überein. Eine schwache Anisotropie der Oberflächen-Energie kann durch das Einsetzen der effektiven Steifigkeit der Fest-Flüssig-Grenzfläche in die Mullins-Sekerka-Theorie berücksichtigt werden. Für den Fall von ternären Systemen mit unterschiedlichen Diffusivitäten wird eine modifizierte Stabilitäts-Theorie hergeleitet. Der Zusatz von Magnesium verschiebt die Stabilitätsgrenzen gegenüber dem rein binären Al-Si System in der selben Weise wie es auch die modifizierte Theorie vorhersagt. Ohne diffusive Wechselwirkungen stimmt die hergeleitete Theorie mit den Ergebnissen von Simulationen mit verschiedenen Einträgen auf der Diagonalen der Diffusivitäten-Matrix überein. Als ein weiterer Punkt werden die Abweichungen zwischen den Simulationen und der Theorie aufgrund von nicht-diagonalen Einträgen der Diffusionsmatrix diskutiert. Für Bedingungen wie eine variierte Stärke der Oberflächen-Energie-Anisotropie oder verschiedene Anteile der dritten Komponente Magnesium ergeben die Simulationen vielfältige Wachstumsmuster, die von „Seetang-Mustern“ über zelluläres Wachstum bis hin zu der Ausbildung von kolumnaren dendritischen Fronten reichen.

Der Einfluss verschiedener Stärken der Oberflächen-Energie-Anisotropie wird ebenfalls für äquiaxiales dendritisches Wachstum der aluminiumreichen FCC-Phase untersucht. Bei einer starken Anisotropie passen die resultierenden Geschwindigkeiten und Radien der Dendriten-Spitzen genau zu der analytischen Lösung für parabolische Platten. Für geringere Stärken der Oberflächen-Energie-Anisotropie verschieben sich die Ergebnisse hin zu der analytischen Lösung für Rotationsparaboloide. Als letztes Thema dieser Arbeit wird ein Konzept für die Kopplung kinetischer Datenbanken mit Phasenfeld-Simulationen am Beispiel der Vergröberung im System Eisen-Kupfer vorgestellt. Die Verwendung solcher Datenbanken führt zu quantitativeren Ergebnissen bei Festkörper-Simulationen, in denen diffusive Prozesse eine entscheidende Rolle spielen.

Abstract

The utilization of thermodynamic and mobility data plays a major role in quantitative phase-field modeling. The present thesis discusses this topic on a theoretical level and also deals with the practical application of such data for different phase-field studies. At first, the basics of thermodynamics and kinetics are presented with a focus on different models for Gibbs energies. These quantities are essential in the field of computational thermodynamics, which gets described together with the approach of computational kinetics. An introduction is given to the quantitative phase-field model based on grand potentials, which is applied for the presented phase-field studies. As these simulations only cover a small composition and temperature range, they can be optimized computationally by the use of simplified expressions for the thermodynamic functions. Different strategies to model these functions are laid out and applied for the specific requirements of certain material systems, which include for example stoichiometric or pseudobinary phases. The usage of Taylor expansions and the least-squares method is discussed regarding the deviations from the original formulations.

This thesis includes a stability analysis of single-phase growth fronts for the Al-Si alloy in dependence of different material and process parameters. For a system of two components having isotropic surface energies, the stability of planar fronts is studied. The simulated growth rates of sinusoidal perturbations match well with the Mullins-Sekerka theory. A weak anisotropy of the surface energy, can be accounted for by inserting the effective stiffness of the solid-liquid interface into the Mullins-Sekerka theory. For the case of ternaries with different diffusivities, a modified stability theory is presented. The addition of magnesium shifts the stability thresholds with respect to the pure binary aluminum-silicon in the same way as the modified theory predicts it. Without diffusional interaction, the derived theory matches with the results of simulations for different entries on the diagonal of the diffusivity matrix. As a further point, the deviations between simulations and theory due to off-diagonal entries in the diffusivity matrix are discussed. For conditions like a varying strength of the surface energy anisotropy or different amounts of the third component magnesium, the simulations yield various forms of growth ranging from seaweed patterns to cellular growth up to the onset of columnar dendritic fronts.

The influence of different strengths of the surface energy anisotropy is also studied for equiaxed dendritic growth of the aluminum-rich FCC phase. With a strong anisotropy, the resulting dendrite tip velocities and radii match closely to the analytical solution for parabolic plates. For lower strengths of the surface energy anisotropy, the results are shifted towards the analytical solution for paraboloids of revolution. As the last topic of this thesis, a concept for the coupling of kinetic databases with phase-field simulations is presented at the example of coarsening in Fe-Cu. The use of such databases yields more quantitative results for solid state simulations, in which diffusional processes play a major role.

Danksagung

Zuerst möchte ich mich herzlichst bei Professor Britta Nestler für die Betreuung meiner Arbeit bedanken und dafür, dass ich auch als Absolvent einer Fachhochschule die Möglichkeit zu promovieren bekommen habe und über viele Jahre am IMP und IAM beschäftigt sein durfte. Ich möchte mich außerdem bei Professor Hans Jürgen Seifert für die Übernahme des Korreferententums und das Bereitstellen von thermodynamischen Datensätzen bedanken. In besonderem Maße gebührt mein Dank meinem langjährigen Betreuer Professor Abhik Choudhury für seine Unterstützung, für zahlreiche Diskussionen und für wichtige Impulse. Das Meiste was ich über Phasenfeld-Modellierung und Erstarrung gelernt habe, habe ich von ihm gelernt.

Mein Dank geht weiterhin an alle Entwickler des Pace3D-Projekts, für die Weiterentwicklung des Solvers und das Bereitstellen vieler nützlicher Tools. Vielen Dank möchte ich im Speziellen Michael Selzer für seine Unterstützung bei der Umsetzung der CALPHAD-Kopplung sagen. Ein besonderes Dankeschön für das Korrekturlesen meiner Dissertation möchte ich Avisor Bhattacharya, Nani Enugala, Michael Kellner, Arnab Mukherjee und Rajdip Mukherjee aussprechen. Ich bin dankbar unzählige Stunden auf den institutseigenen Clustern gerechnet haben zu können. Für alle technische Unterstützung gilt mein Dank den Systemadministratoren Christof Ratz, Jonathan Buch und Andreas Müller. Vielen Dank an sämtliche Gruppenmitglieder. Ich bin dankbar in der gesamten Zeit ständig eine große Hilfsbereitschaft, Freundlichkeit und fachliche Kompetenz erfahren zu haben. Auch wenn ich an dieser Stelle nicht jeden/jede aufzähle, so bin ich doch froh euch alle kennen gelernt zu haben.

Desweiteren bin ich der Europäischen Union, dem Land Baden-Württemberg und der Deutschen Forschungsgemeinschaft zu Dank für die finanzielle Unterstützung meiner Arbeit verpflichtet.

Zu guter Letzt gilt meine tiefe Dankbarkeit meinen Eltern, meiner Familie und all meinen Freunden. Danke dass ihr mich unterstützt und immer wieder für Ablenkung und Aufheiterung gesorgt habt.

Contents

I Introduction

1 Motivation 5

2 Outline 7

II Background

3 Fundamentals of thermodynamics 11
3.1 Thermodynamic relations 11
3.2 Models for the Gibbs energy 15
3.2.1 Gibbs energy for phases with fixed compositions 16
3.2.2 Gibbs energy for solution phases 17
3.3 Phase diagrams 23
3.4 Binary systems 23
3.4.1 Gibbs phase rule 27
3.5 Ternary systems 29
3.6 Vertical sections 30

4 Fundamentals of diffusion 35

5 Computational thermodynamics and kinetics 39
5.1 CALPHAD method 40
5.1.1 Data assessment 41
5.1.2 Gibbs energy formulation 41

5.1.3 Compound-energy formalism 42
5.1.4 Simplified formulation for the binary case 43
5.1.5 Ternary contribution of the excess part 45
5.1.6 Parameters dependent on temperature 46
5.1.7 Modeling of stoichiometric phases in CALPHAD 47
5.1.8 Magnetic contribution in the CALPHAD method 48
5.1.9 Further models 49
5.2 Computational Kinetics 50

6 Analytical models for solidification 53
6.1 Mullins Sekerka theory 53
6.2 LGK theory 57

7 Description of the phase-field model 59
7.1 General phase-field concepts 60
7.2 Grand potential model 63
7.2.1 Definitions used in the grand potential model 63
7.2.2 Temperature dependence of the chemical potential . . . 74
7.2.3 Boundary conditions 76

III Coupling of the grand potential model with CALPHAD databases

8 Overview of coupling methods 81

9 Overview of system setting data 83

10 Simplified formulations for the thermodynamic functions . . . 87
10.1 Formulation as quadratic polynomials 87
10.2 Formulation for the ideal solution model 90

11 Determination of model parameters from CALPHAD data . . 93
11.1 Taylor expansion 93
11.1.1 Approximation of the coefficients of the polynomial model 94
11.1.2 Approximation of the coefficients of the ideal solution model 97

11.2 Approximating the thermodynamic functions with the least squares method 99
11.3 Models for the temperature dependence 100
11.4 Modeling approach for stoichiometric phases 102
11.5 Modeling pseudobinary phases 108

12 The trade-off between accuracy and complexity 111
12.1 Computational effort 111
12.2 Analysis of the deviation arising from a quadratic fit 112
12.2.1 Deviations due to Taylor approximations of the ideal part 113
12.2.2 Least squares fitting of the ideal part 119

13 Conclusions and Outlook 123

IV Phase-field studies based on CALPHAD data

14 Stability analysis of single-phase growth in Al-Si-(Mg) alloys . 129
14.1 Thermodynamic functions for the binary simulations 131
14.2 Mullins Sekerka study of perturbed growth fronts 136
14.2.1 Simulation study for isotropic and anisotropic surface energies 136
14.2.2 Simulation study of ternary impurity 146
14.2.3 Conclusions 157
14.3 Phase-field simulation of equiaxed dendritic growth 159
14.3.1 Simulation setup 159
14.3.2 Comparison to analytical predictions from LGK theory 159
14.3.3 Conclusions 163

15 Ostwald ripening of Fe-Cu - Influence of diffusion coefficients 165
15.1 Simulation setup based on data from computational thermodynamics and kinetics 166
15.2 Simulation series for different diffusivities 171
15.3 Conclusions 175

V Conclusions and Outlook

16 Conclusions **179**
16.1 CALPHAD coupling 179
16.2 Phase-field simulations 181

17 Outlook **183**

VI Appendix

A Deviation analysis for Gibbs energy approximations **187**

B Deviation analysis for approximations of the phase concentrations **191**

C Proof of the invertibility of the chemical potential functions of the Al-Si system **195**

Bibliography **199**

List of symbols

Abbreviations and Acronyms

BBM	Broken Bond Model
BCC	Body-Centered Cubic
CALPHAD	CALculation of PHase Diagrams
CEF	Compound-Energy Formalism
FCC	Face-Centered Cubic
ICME	Integrated Computational Materials Engineering
KMC	Kinetic Monte Carlo method
LGK	Lipton-Glicksman-Kurz
LSW	Lifshitz-Slyozov-Wagner Law
MS	Mullins Sekerka
PACE 3D	Parallel Algorithms for Crystal Evolution in 3D
PF	Phase-Field
SER	Standard Element Reference state
TDB	Thermodynamic DataBase

Indices

α, β, γ	Phase indices
A, B, C, K, i, j, k	Component indices
F	Approximated by least squares fitting
I	Ideal solution approximation
m	Molar quantity
P	Approximated with a polynomial formulation
s	Sublattice index
T	Approximated by Taylor expansion

Vectors and matrices

$\mathbf{c}$	Vector of concentrations
$\mathbf{c}^\alpha$	Vector of phase concentrations
$\mathbf{c}_0$	Composition vector for Taylor expansion
$\mathbf{H}_{f^\alpha}(\mathbf{c})$	Hessian matrix for function f^α
$\boldsymbol{\mu}$	Vector of chemical potentials
$\boldsymbol{\mu}_{eq}$	Vector of equilibrium chemical potentials
$\boldsymbol{\mu}^0$	Vector of macroscopic interfacial chemical potentials in the sharp interface limit
$\boldsymbol{\phi}$	Vector of phase-field parameters
$\mathbf{x}$	Vector including the mole fractions
$\boldsymbol{\xi}^\alpha$	Coefficient vector for polynomial approximation
$\boldsymbol{\Xi}^\alpha$	Coefficient matrix for polynomial approximation

Roman symbols

$a(\phi, \nabla\phi)$	Gradient energy function	$[\mathrm{J/m^4}]$
a_s	Number of sites of sublattice s	$[-]$
b_i	Stoichiometry factor	$[-]$
C	Heat capacity	$[\mathrm{J/K}]$
c, c_i, c_{eq}, c^*	Concentration (, of component i, at equilibrium, certain concentration)	$[-]$ in PF model, $[\mathrm{mol/m^3}]$ else
c_i^α,	Phase concentration of component i and phase α	$[-]$
c_{id}	Phase concentration corresponding to the ideal part of the Gibbs energy	$[-]$
C_p	Heat capacity at constant pressure	$[\mathrm{J/K}]$
C_V	Heat capacity at constant volume	$[\mathrm{J/K}]$
c_0	Liquid concentration at the interface, concentration around which a Taylor expansion is done	$[-]$
C_1	Concentration amplitude of a perturbed front	$[-]$
c_∞^s	Solute conc. in matrix at flat interface	$[-]$

c_∞	Liquid concentration far away from the front	[–]
D	Constant for the calculation of the magnetic part of the Gibbs energy	[–]
D, D_{ij}, D^{liq}	Coefficient for interdiffusion (, between component i and j, in the liquid phase)	[m^2/s]
D_i^*	Tracer diffusion of component i	[m^2/s]
d	Number of dimensions	[–]
d_0	Capillary length	[m]
F	Helmholtz (free) energy, degrees of freedom	[J], [–]
$\tilde{F}$	Solvability Integral	[–]
$f, f^\alpha, f_A^\alpha, f_{AB}^\alpha$	Free energy density (, of phase α) and parameters for its calculation	[J/m^3]
${}^{red}f^\alpha(T,\mathbf{c})$	Reduced function for the free energy densities of phase α	[J/m^3]
f_α	Phase fraction of phase α	[–]
G, G'	Thermal gradients	[K/m]
$G, \Delta G_{mix}, G_m^\alpha$	Gibbs (free) energy (, of mixing, of phase α referred to one mole of a substance)	[J] (, [J], [J/mol])
${}^\circ G_i^\alpha, {}^\circ G_{I_0}^\alpha$	Gibbs energy of formation for pure component i, for constituent array I_0	[J], [J/mol]
${}^{ex}G_m^\alpha$	Excess part of Gibbs energy	[J/mol]
${}^{id}G_m^\alpha$	Ideal mixing part of Gibbs energy	[J/mol]
${}^{mag}G_m^\alpha$	Magnetic part of Gibbs energy	[J/mol]
${}^{phys}G_m^\alpha$	Additional part of Gibbs energy	[J/mol]
${}^{ref}G_m^\alpha$	Reference part of Gibbs energy	[J/mol]
$\overline{G}$	Gibbs energy of a two-phase mixture	[J]
G_c	Gradient of concentration field	[1/m]
$g_\alpha(\phi)$	Interpolation function for phase α (to interpolate mobilities)	[–]
$g(\tau)$	Function to calculate magnetic part of Gibbs energy	[–]
$g(\omega)$	Function to calculate rates of change of a perturbed interface	[–]
$H, \Delta H_{mix}$	Enthalpy (, of mixing)	[J]

H_i^{SER}	Enthalpy of constituent i at the standard element reference state	[J]
$h_\alpha(\phi)$	Interpolation function for phase α	[−]
$I_k, I_0, I_1, \overline{I_0}, \overline{I_1}$	Constituent array of order k, intervals	[−], [−]
$I(\mathrm{Pe})$	Ivantsov integral	[−]
J_i	Flux of component i	[m/s]
$\vec{j}_i$	Anti-trapping current	[m/s]
K	Number of components	[−]
K_{LSW}	Coarsening rate constant	[m^3/s]
k	Partition coefficient $k = c_{eq.}^{solid}/c_{eq.}^{liquid}$	[−]
k_ω	Decay rate for frequency ω	[1/m]
${}^{v}L_{A,B}^{\alpha}(T),\ {}^{i}L_{ijk}^{\alpha}$	Interaction parameters for Gibbs energy calculation	[J/mol]
l_D	Solutal diffusion length	[m]
L_{ij}, L'_{ij}	Phenomenological parameters for diffusivity calculation	[mol^2/mJs]
L_1, L_2	Liquidus curves	[−]
$\tilde{M}$	Solvability Integral	[−]
M_A°	Frequency factor for mobility calculation	[m^2/s]
M_i	Atomic mobility of component i	[m^2mol/Js]
M_{ij}^α	Mobilities in grand potential model	[m^5/Js]
m	Liquidus slope	[K]
N, N_i	Amount of substance (, of comp. i)	[mol]
n	Index of a timestep	[−]
n	Number of sublattices	[−]
n_s	Number of constituents in sublattice s	[−]
P	Number of phases	[−]
$P_{I_k}(Y)$	Function, which returns the product of all nonzero site fractions in Y	[−]
Pe	Péclet number	[−]
p	Pressure, crystallography constant, partition coef. $p = 1 - c_{eq.}^{solid}/c_{eq.}^{liquid}$	[Pa], [−], [−]
Q	Heat	[J]
Q_A	Activation enthalpy	[J/mol]
$q_{\alpha\beta}$	Normal vector to the interface	[1/m]
R	Ideal gas constant	[J/Kmol]
r, r_{avg}	Radius, average radius	[m]

S, ΔS_{mix}	Entropy (, of mixing)	[J/K]
S_1, S_2	Solidus curves	[–]
T, T^*	Temperature (, certain temperature)	[K]
T_C, $T_{C,i}$, $T^{\nu}_{C,AB}$	Critical temperature and parameters for its calculation	[K]
T_e	Eutectoid temperature	[K]
T_m	Melting temperature	[K]
T_0	Base temperature of temp. gradient	[K]
t	Time	[s]
$\bar{t}$	Non-dimensional time	[–]
U	Internal energy	[J]
V	Volume	[m^3]
V_m, V_j	Molar volume, partial molar volume of component j	[m^3/mol]
v	Velocity	[m/s]
$w(\phi)$	Energy potential function	[J/m^2]
x	Spatial coordinate	[–]
y	Spatial coordinate	[–]
y^s_i	Site fractions in sublattice s	[–]
Y	Matrix containing the site fractions	[–]
Z	Arbitrary physical property	[–]
z	Spatial coordinate	[–]
z_0	Offset of temperature gradient in z-direction	[m]

Greek symbols

β, β_i, β^{ν}_{AB}	Average value of magnetic moment and parameters for its calculation	[–]
Γ	Gibbs-Thomson Coefficient	[Km]
Δc	Difference of solid-liquid equilibrium concentrations	[–]
Δt	Time step width	[s]
Δx	Cell size	[m]
δ	Amplitude	[m]
$\delta_{\alpha\beta}$	Strength of the surface energy anisotropy	[–]

δ_{ij}	Kronecker symbol	$[-]$
ε	Parameter related to the width of the interface, half of the width of a concentration range	[m], $[-]$
κ	Curvature	$[-]$
κ_B	Boltzmann constant	$[J/K]$
λ	Lagrange Parameter in PF model	$[J/m^3]$
λ, λ_0	Wavelength, critical wavelength	[m]
$\overline{\lambda}$	Mean spacing	[m]
μ, μ_i, μ_{eq}	Chemical potential (, of component i, at equilibrium)	$[J/m^3]$ in PF model, $[J/mol]$ in general
μ_i^∞	Chemical potential of component i far away from the front	$[J/m^3]$
v_i	Fraction used to calculate ternary Gibbs energy	$[-]$
σ, $\sigma_{\alpha\beta}$	Surface energy (, between phase α and β)	$[J/m^2]$
σ_{eff}	Effective surface energy (effective interface stiffness)	$[J/m^2]$
σ^*	Stability constant used in LGK analysis	$[-]$
τ	Nondimensional temperature	$[-]$
τ, $\tau^{\alpha\beta}$	Interface relaxation parameter (, for α-β-interface)	$[Js/m^4]$
Φ_A, Φ_A^i, ${}^v\Phi_A^{i,j}$	Parameters for mobility calculation	$[J/mol]$ or $[\ln(m^2/s)]$
ϕ_α	Order parameter of phase α	$[-]$
ϕ_α^0	Solution of the phase-field equation in the asymptotic analysis	$[-]$
Ψ	Grand potential density in PF model, grand potential else	$[J/m^3]$ in PF model, [J] else
Ω	Grand Potential Functional	[J]
$\Omega_{AB}(T)$	Interaction parameter to calculate the enthalpy of mixing	[J]
Ω_{liquid}, Ω_{solid}	Liquid and solid part of a domain	$[-]$
ω, ω_0	Frequency, critical frequency	$[1/m]$

Part I

Introduction

The manufacturing process of casting has been used by mankind since many centuries and is still the most important method to bring raw metal into a first shape. By subsequent working steps the cast can be transformed into the final product and its properties can be further improved. However, the quality of the final good, to a large extent depends on the initial casting process. The closer the workpiece resembles the final shape after the stage of casting and the better its material properties already are at that point, the less additional working steps are necessary. This in turn lowers the costs of production, which explains the particular importance to better understand and optimize that key process.

Since centuries the empirical knowledge about the secrets of metal processing has been passed on from masters to their apprentices. In the course of industrialization with the scientific revolution happening parallel, the inner structure of metals and alloys together with the physical processes causing their formation have become a scientific object, which is treated in the discipline of metallurgy. It was found, that some parts of the microstructure can be completely regular and periodic, like it is the case in crystallines, while in other regions and on other length scales the materials can be characterized by complex patterns. The invention of various instruments enabled an insight into these microstructures and helped to understand the underlying mechanisms of their evolution. The methods of analysis have been constantly improved and today a whole spectrum of techniques is available for specific cases of application, including light microscopy, electron microscopy, X-ray analysis and many more.

Further progress has also been made in theoretical physics and the branch of thermodynamics, which led to analytical models of solidification. The provided equations are used by metallurgists as mathematical tools to optimize the process parameters. However, all of these analytical solutions are based on simplifications and describe the real physics only partially. With the advent of the information age, more precise techniques for numerical predictions and optimizations became feasible. Accompanied by the increase of computer power, different simulation approaches for the modeling of material processes and microstructure formation emerged. Hereby, the choice of the appropriate simulation technique is dependent on the length scale of the treated problem. For example, the layout of the mold together with the process parameters can be optimized on a macroscopic level with the help of casting process simulations. On the microstructural scale, which has a significant influence on the material properties, the phase-field method proved to be a useful approach [1]. Since the pioneering work in the previous century, the phase-field method has become a wide-spread simulation model, which is nowadays applied in many other fields of application apart from solidification.

However, accurate data about the thermodynamic properties of the material system is a prerequisite to achieve realistic results, which can be derived from the thermodynamic databases provided by the CALPHAD method [2]. In this approach, the Gibbs energies are used as representative functions, containing the necessary information to calculate phase diagrams. For this purpose, suitable formalisms have been developed, which describe the Gibbs energies over the whole concentration range and in the temperature regime of technical interest. For the application in simulations, however, more often than not it is convenient to approximate the required thermodynamic functions in the region of interest. Given these input parameters, the results from the phase-field method are found to be in good agreement with analytical predictions for basic setups. In addition, realistic results can be achieved for larger and more complex systems and the current impediments for a broader application of material simulations are still due to the availability of computational resources. If the development of more powerful processing units continues at the same rates as in the past decades, these limitations can be overcome in the near future and the cost savings of simulations compared to experiments will become more and more notable.

1 Motivation

The work presented in this PhD thesis was carried out as part of the "Center of Computational Materials Science and Engineering (CCMSE)", which was a joint research project of different universities in the German state of Baden-Württemberg. The aim of this project was to make progress in the interdisciplinary field of computer-aided material science and one of the investigated subjects was the formation of microstructures in casting processes. Because aluminum-silicon is an industrially relevant non-ferrous alloy system with excellent casting properties [3, 4, 5], it was chosen as an exemplary object of investigation. The particular purpose of the present PhD thesis is to simulate the solidification of this alloy by utilizing thermodynamic data provided by the CALPHAD method [6]. As part of the same project, the phase-field model based on the grand potential formulation of Choudhury and Nestler [7] was developed and implemented simultaneously to my doctoral studies. The core of this model are thermodynamic functions and the key to ensure their quantitativeness is the utilization of accurate data. For this reason, a large part of the present thesis is about the coupling with thermodynamic databases to provide the specific input parameters needed for the grand potential model. This thesis is intended to give an overview about possible coupling approaches and to analyze the different strategies theoretically and at the example of real systems. On the basis of the discussed coupling framework, the solidification of Al-Si under different conditions is investigated, proving the capability of the new phase-field model to cope with real alloy systems. As a validation of the model and its implementation, the simulation studies are designed for the comparison with analytical solutions, such as the well established theories of Mullins and Sekerka [8] or Lipton, Glicksman and Kurz [9]. As a further application, the discussed simulation framework is applied to study the diffusion controlled process of Ostwald ripening in a solid iron-copper alloy.

Over the last years, the new field of Integrated Computational Materials Engineering (ICME) emerged and aroused the interest of both the scientific community and the industry. This international effort deals with establishing of standards and interfaces to link the various methods used in materials science and to benefit from the opportunities of the Digital Revolution. The overall aim behind ICME is to enable the cost-efficient design and optimization of new materials and processes, which is crucial for further technical progress in a world of limited resources. This thesis acts as a small contribution to the ICME effort and is motivated by the same objectives. An overview about the covered topics is given in the following.

2 Outline

At the beginning of this thesis (chapter 3) and 4, the thermodynamic and diffusional principles, which are underlying the processes of structure formation, are described. Next, a brief explanation of the CALPHAD method and computational kinetics (chapter 5.1 and 5.2) follow. In chapter 6, analytical solutions for solidification are described and chapter 7 provides a short introduction to the phase-field model based on grand potentials. Then, different approaches for the coupling of thermodynamic datasets with the phase-field model are described in III. This is followed by the application of the coupling framework to solidification simulations of Al-Si-Mg alloys. A series of simulations deals with the effect of surface energy anisotropy on the formation of different growth morphologies starting from planar fronts 14.2.1. As a further point, the addition of a third component in combination with different diffusivities is analogously investigated in 14.2.2. Chapter 14.3 is about equiaxed dendritic solidification of Al-Si for different strengths of anisotropy and the comparison with analytical solutions. Finally, a study about Ostwald ripening of Fe-Cu demonstrates the utilization of kinetic datasets in phase-field simulations (chapter 15). The thesis concludes with a discussion and summary of the results together with an outlook about future opportunities V.

Part II

Background

3 Fundamentals of thermodynamics

During the 17th and 18th century the groundwork for thermodynamics was laid with the development of the thermometer and elementary studies on the principal mechanisms from this field of research. The formulation of the four laws of thermodynamics and the development of the fundamental thermodynamic equations in the 19th and early 20th century set this new discipline onto a scientifically profound basis. For instance Ludwig Boltzmann found a microscopic explanation for the previously rather notional concept of entropy and founded the branch of statistical mechanics. Another person, whose name is inextricably linked with the field of thermodynamics is Josiah Willard Gibbs. He made pioneering work in physical chemistry and introduced key concepts like the chemical potential [10], the Gibbs free energy and the Gibbs phase rule, which are often used throughout this thesis. In the following paragraphs, the basic thermodynamic principles are explained, which are required in the context of the CALPHAD and the phase-field method.

3.1 Thermodynamic relations

Thermodynamics deal on a macroscopic level with the various manifestations of energy and the resulting processes without making statements about their speed. In the context of metallurgy, thermodynamics can be used to determine the equilibrium states of material systems, which can correspond to diverse configurations of the components depending on the prevailing conditions. At equilibrium there is no driving force for a transformation of the system, even though the properties in different phases might not be the same. The state of a thermodynamic system can be described by the variables for its macroscopic properties. If these state variables are independent of the system size, they are called intensive variables, such as temperature T or pressure p. Extensive state variables, like volume V or the amount of substance N are dependent on the size of the system. Another extensive variable is the internal energy U. The internal energy arises from the internal forces of a system acting between its particles. It consists of the energy due to the translational, rotational and vibrational motion of the particles and also of their rest mass energy. The internal energy only describes the energy contained in a system and does not include the kinetic and potential energy of the system as a whole in relation to its surrounding. Another thermodynamic quantity is the enthalpy

$$H = U + pV. \tag{3.1}$$

This form of energy can be thought of as the sum of the internal energy U of a body and the work pV, that had to be done by the system to introduce the body of volume V into space. Compared to the internal energy the term pV is in general small for solids and liquids, such that U is the dominating part of the enthalpy [11]. The temperature dependence of the enthalpy and the internal energy is connected to the heat capacity C of a system. This quantity gives a proportionality between the heat Q, that is added to a system and the temperature change related to it:

$$C = \left(\frac{\partial Q}{\partial T}\right). \tag{3.2}$$

The derivative of the internal energy with respect to temperature gives the heat capacity at constant volume

$$C_V = \left(\frac{\partial U}{\partial T}\right)_V, \tag{3.3}$$

whereas the heat capacity for constant pressure is given as a derivative of the enthalpy

$$C_p = \left(\frac{\partial H}{\partial T}\right)_p . \tag{3.4}$$

These relations are important for the assessment of thermodynamic data explained in section 5.1.1.

Both, the internal energy and the enthalpy are thermodynamic potentials. Like the gravitational potential defines, whether a body changes its position to get to a lower potential, thermodynamic potentials are quantitative measures for the tendency of transformation of a substance. These quantities have a minimum at states, which are in thermodynamic equilibrium. This is true under the condition, that their characteristic variables are kept constant. For a closed isothermal system of constant volume, equilibrium conditions are characterized by a minimum of the Helmholtz free energy

$$F = U - TS \tag{3.5}$$

with S as the entropy. The Helmholtz free energy is also known as Helmholtz energy or just free energy. The adjective "free" is used because the change of the Helmholtz free energy during an isothermal process corresponds to the available work, that can be done by the system.

The Helmholtz free energy relates the internal energy to the entropy S. In statistical mechanics the entropy is defined as a measure for the statistical probability of a macroscopic state. Hereby a macroscopic state is given by the same values of the macroscopic variables such as temperature, density or the internal energy. In contrast, the microstate of a system is clearly defined by the positions and momentums of all its particles. One and the same macroscopic state can be the manifestation of a variety of energetically equivalent microstates. The larger the number of equivalent microstates, the bigger is the probability of the corresponding macroscopic state and its entropy. Thus, for a system with high entropy the actual microscopic state is highly uncertain. According to the second law of thermodynamics, spontaneous changes in an isolated system take place, such that the entropy of the system increases.

While the Helmholtz free energy is related to processes with changing pressure, for most experiments the pressure and temperature are known and hence the appropriate thermodynamic potential for these kind of processes is the Gibbs free energy G. For the sake of brevity and in consistence with many publications about thermodynamics and the CALPHAD method [12, 13], this quantity is referred to as Gibbs energy hereinafter. The Gibbs energy is defined as the difference

$$G = H - TS. \tag{3.6}$$

For isothermal and isobaric conditions without mass transfer, phase transitions take place if they result in a reduction of G. Due to the definition of the Gibbs energy, the stability of a phase for lower temperatures mainly depends on the enthalpy, whereas the entropy is dominating for higher temperatures. Thus, gases with high entropy are stable at high temperatures, while solid phases with a small enthalpy are found at low temperatures. A thermodynamic potential for open systems is the Grand (or Landau) potential, which is defined as

$$\Psi = U - TS - \sum_{i=1}^{K-1} \mu_i N_i = F - \sum_{i=1}^{K-1} \mu_i N_i. \tag{3.7}$$

At a minimum of Ψ an isothermal, isochoric system with constant chemical potentials is in thermodynamic equilibrium. Such conditions are given for a system with a fixed volume, that can exchange energy and mass with a large reservoir system to stay in thermal and chemical equilibrium with it. To describe the energetic differences connected to an exchange of particles J. W. Gibbs introduced the concept of chemical potentials. They are defined as the partial derivative of G with respect to the amount of substance of a component i

$$\mu_i = \left(\frac{\partial G}{\partial N_i} \right)_{T,p,N_{j \neq i}} \tag{3.8}$$

under the condition of constant temperature, pressure and with constant amounts of all other constituents N_j. An explanation of this definition is given by the schematic drawing in figure 3.1. If one atom of type A is added to a system and the number of all other atoms is conserved, the total number of atoms changes. This addition increases the Gibbs energy of the whole system by dG. For an addition of a small number of atoms dN_A the energy changes linearly and the proportionality factor is given by the chemical potential $dG = \mu_A dN_A$. The chemical potential is thus related to the work, that is required to keep a system in thermodynamical equilibrium, if one particle is added to it.

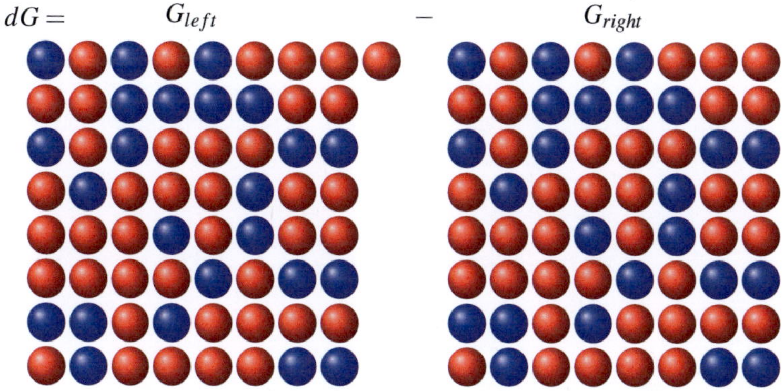

Figure 3.1: Visualization of the definition of the chemical potential. The right side shows the original system consisting of atoms A (red) and B (blue). On the left side one additional atom of A is added to the system. The change in Gibbs energy dG due to this addition is given by the difference between the total Gibbs energy of the system on the left and the one of the original system.

3.2 Models for the Gibbs energy

As pointed out before, the Gibbs energy is an essential quantity for phase transformations and it is important to describe it in an appropriate manner. This thermodynamic quantity plays a central role in the CALPHAD method and in different phase-field models. As the utilization of the Gibbs energies from thermodynamic databases is a principal topic of this thesis, the most common models for their description are explained in the following. Furthermore the basic principles of phase-diagram calculation based on Gibbs energies are addressed in the subsequent sections.

3.2.1 Gibbs energy for phases with fixed compositions

All models for higher order systems in the CALPHAD method, which are explained in section 5.1.2, refer to the Gibbs energies of the unary systems. These basic systems consist of only one element and hence have a fixed composition. But also stoichiometric phases are independent of composition, as the ratio of their different components is explicitly defined. Since most phase transformations of metals and alloys occurring in industrial processes and especially the ones considered in this thesis happen at the constant atmospheric pressure, the formulation for the Gibbs energies only have to be dependent on temperature and can be written as

$$G(T) = a + bT^{-1} + cT\ln(T) + \sum_{n=1} d_n T^n. \tag{3.9}$$

The coefficients $a, b, c, \cdots$ can be adjusted to fit the temperature dependence obtained from experiments, which is explained in section 5.1.1. For such a formulation, the other fundamental thermodynamic functions can be easily derived from G, such as the enthalpy, the entropy and the heat capacity at constant pressure:

$$H(T) = G(T) - T\left(\frac{\partial G(T)}{\partial T}\right)_{p,N_i} = a + 2bT^{-1} - cT - \sum_{n=1}(n-1)d_n T^n, \tag{3.10}$$

$$S(T) = -\left(\frac{\partial G(T)}{\partial T}\right)_{p,N_i} = bT^{-2} - c - c\ln(T) - \sum_{n=1} n d_n T^{n-1}, \tag{3.11}$$

$$C_p(T) = -T\left(\frac{\partial^2 G(T)}{\partial T^2}\right)_{p,N_i} = -c - 2bT^{-2} - \sum_{n=1} n(n-1) d_n T^{n-1}. \tag{3.12}$$

3.2.2 Gibbs energy for solution phases

For solution phases, the composition is not fixed anymore and has to be taken into account. In the following, the common models for solution phases are introduced, whereby the equations are formulated for systems with only two components A and B. The simplest type of a composition dependent formulation is the energy of mechanical mixture. The Gibbs energy for such a mixture gets calculated as

$$G^{\alpha}(T, x_A, x_B) = {}^{\circ}G^{\alpha}_{A}(T)x_A + {}^{\circ}G^{\alpha}_{B}(T)x_B \tag{3.13}$$

with $x_i = N_i/N$ as the mole fraction of component i. The Gibbs energy for each phase is given as an interpolation between the Gibbs energies ${}^{\circ}G^{\alpha}_{i}(T)$ of the pure components, which can be expressed with a formulation like in equation 3.9. A visualization of the model is shown in figure 3.4. This model is very simplistic as it considers the whole system as a purely mechanical mixture of its parts and does not consider the changes due to interactions between the constituents. The model can be thought of as a set of different constituents, for which the atoms of one sort are separated from the atoms of the other sorts and do not interact with them, as displayed in figure 3.2. If the atoms are mixed randomly like in figure 3.3 and interact with each other, further terms have to be added to take care about these mixing contributions.

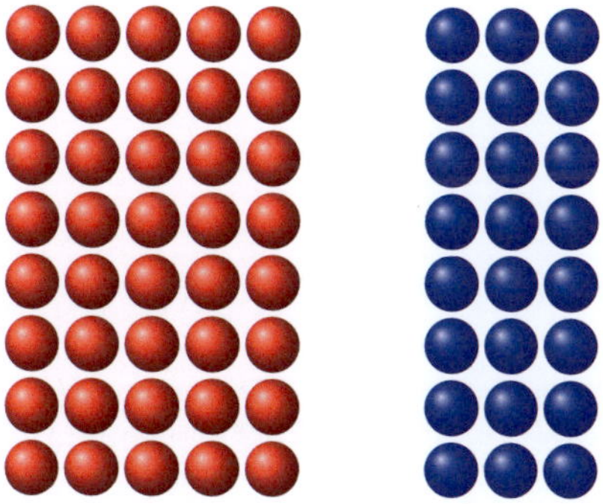

Figure 3.2: Separated atoms of type A and B, representing a purely mechanical mixture.

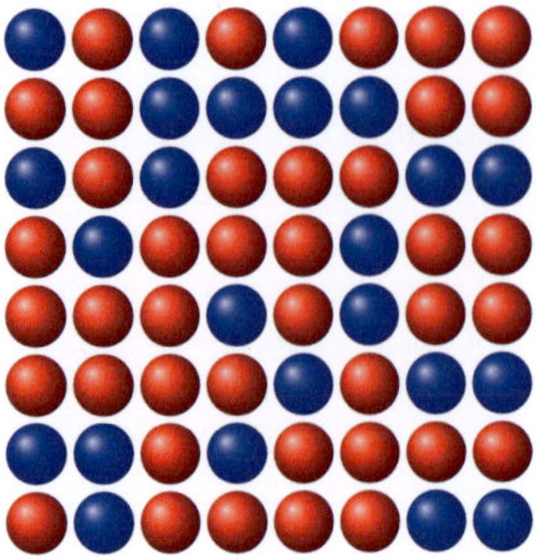

Figure 3.3: Randomly distributed atoms, which have Gibbs energy contributions due to their mixing.

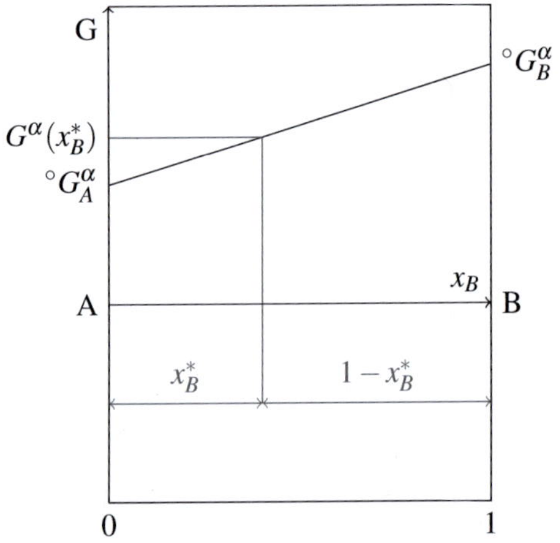

Figure 3.4: Energy of a mechanical mixture: the Gibbs energy of the solution phase α at a certain temperature and for a composition x_B^* gets calculated as a linear interpolation between the energies of the pure components $^\circ G_A^\alpha$ and $^\circ G_B^\alpha$.

The ideal solution model is the simplest type of model including contributions due to mixing. Hereby, the differences to the Gibbs energy of a mechanical mixture are only related to the change in configurational entropy. Given a random distribution of the atoms, the contribution can be derived from statistical thermodynamics in the following way. Let us assume that, a system consists of two components, which are both perfectly soluble in one another at a temperature below the melting point. With N_A atoms of type A and N_B atoms of type B, there are $N = N_A + N_B$ atoms in total. The number of possibilities to arrange these atoms to a crystal (the number of microstates) is $\frac{N!}{N_A!N_B!}$. Inserting this into Boltzmann's entropy formula, one gets:

$$\begin{aligned}\Delta S_{mix}(N, N_A, N_B) &= \kappa_B \ln\left(\frac{N!}{N_A!N_B!}\right)\\ &= \kappa_B\left(\ln(N!) - \ln(N_A!) - \ln(N_B!)\right)\end{aligned} \tag{3.14}$$

with κ_B as the Boltzmann constant. By applying Stirling's approximation $\ln(N!) \approx N\ln(N) - N$ and the constraint $x_A + x_B = 1$, an equation for an approximation of ΔS_{mix} can be written as

$$\Delta S_{mix}(N, x_A, x_B) \approx -\kappa_B N\left(x_A \ln x_A + x_B \ln x_B\right). \tag{3.15}$$

Because x is defined between 0 and 1, the results of the logarithmic functions are negative for any solution, while the entropy of mixing must be positive. The configurational entropy of the pure components is defined as zero and thus the mixing of atoms is always preferred for an ideal solution. By multiplying with $-T$, the Gibbs energy of mixing can be obtained from the entropy of mixing:

$$\begin{aligned}\Delta G_{mix}(T, N, x_A, x_B) &= -T\Delta S_{mix}(T, N, x_A, x_B)\\ &= N\kappa_B T\left(x_A \ln x_A + x_B \ln x_B\right).\end{aligned} \tag{3.16}$$

Since the Gibbs energy of the system is proportional to the number of atoms present, one can replace $N\kappa_B = R$ with R as the gas constant related to one mole and obtain the following equation:

$$\Delta G_{mix}(T,x_A,x_B) = RT\,(x_A \ln(x_A) + x_B \ln(x_B))\,. \tag{3.17}$$

This difference with respective to the purely mechanical mixing ΔG_{mix} is shown in figure 3.5 and the subtraction from the linear interpolation in figure 3.6.

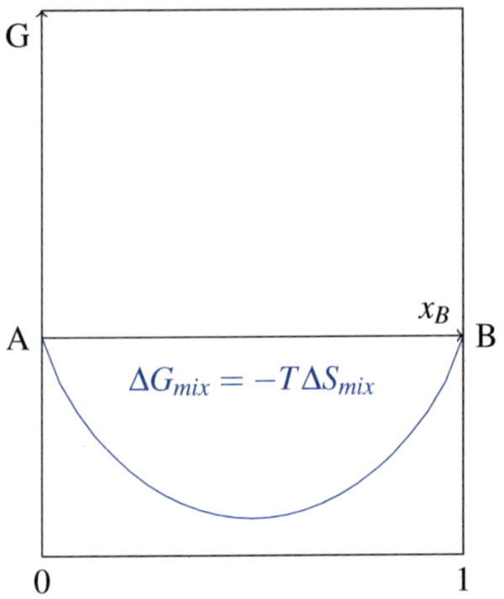

Figure 3.5: The entropy due to mixing ΔS_{mix} leads to a change of the Gibbs energy ΔG_{mix}. This symmetric contribution is always negative or zero and gets minimal for the equiatomic state, in which all components have the same amounts.

For an ideal solution the mixing does not cause a change of energy due to the creation and braking of atomic bonds and though there is no enthalpy of mixing involved. Usually the interchange of atoms happens either endo- or exothermic for a positive or negative enthalpy of mixing, respectively. The regular solution model takes this into account with an additional term

$$\Delta H_{mix}(T,x_A,x_B) = x_A x_B \Omega_{AB}(T), \tag{3.18}$$

which is displayed in figure 3.7. It is a symmetric solution as the interaction between atoms A-B is assumed to be identical to B-A and can be described by a single interaction parameter $\Omega_{AB}(T)$, which is in general dependent on temperature. The product $x_A x_B$ hereby ensures the requirement, that ΔH_{mix} is zero at the terminal compositions (the compositions indicating a pure component).

The simplest type of asymmetric solutions is the sub-regular solution model, for which the interactions between A-B and B-A are different. It is described with the formula

$$\begin{aligned} G^{\alpha}(T,x_A,x_B) = & {}^{\circ}G_A^{\alpha}(T)x_A + {}^{\circ}G_B^{\alpha}(T)x_B + RT\left(x_A \ln(x_A) + x_B \ln(x_B)\right) + \\ & x_A x_B \left(x_A \Omega_{AB}(T) + x_B \Omega_{BA}(T)\right) \end{aligned} \tag{3.19}$$

and displayed in figure 3.8. The subregular solution model may also be considered as a weighted average of two regular solution models and thus it also has the property, that the enthalpy of mixing vanishes at the terminal compositions. To calculate phase diagrams according to the CALPHAD method, more sophisticated Gibbs energy descriptions are required, which are addressed in section 5.1.2.

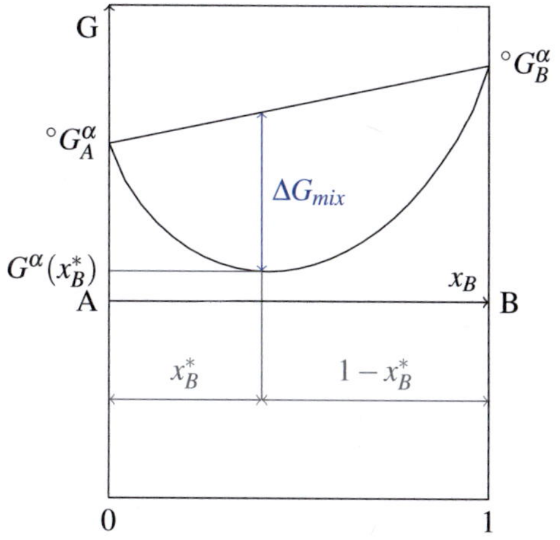

Figure 3.6: For the ideal solution model the additional ΔG_{mix} due to the entropy of mixing is added to the energy of mechanical mixture and lowers the resulting Gibbs energy.

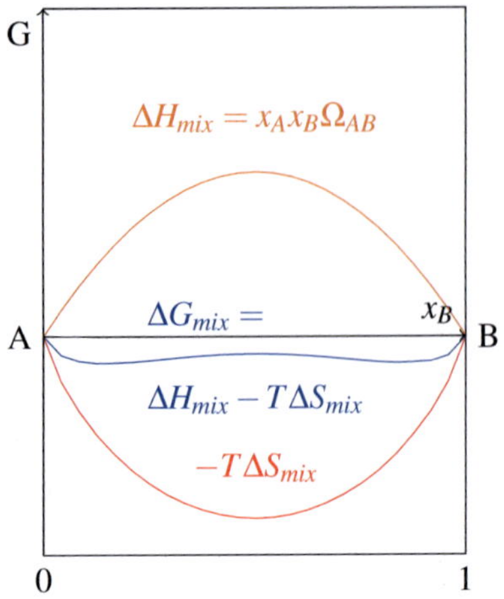

Figure 3.7: The regular solution model includes an additional enthalpy of mixing ΔH_{mix}. In this example ΔH_{mix} is positive, which means, that the mixing is endothermic.

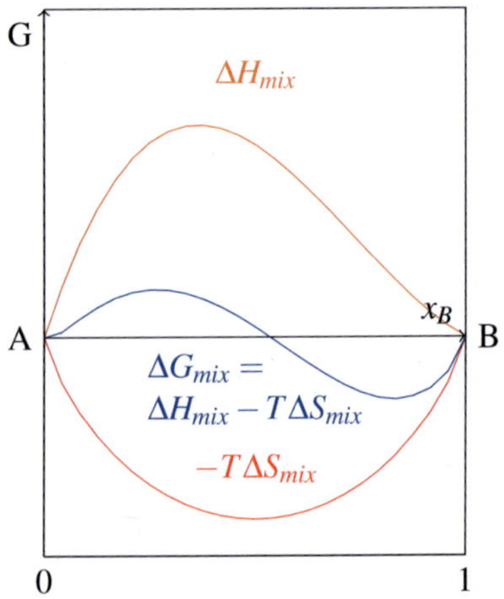

Figure 3.8: For the subregular solution model the enthalpy of mixing is not symmetric anymore and leads to an asymmetric Gibbs energy.

3.3 Phase diagrams

Phase diagrams are essential tools for the study and design of materials. They visually describe the equilibrium constitution, which a material system of a certain composition takes, in dependence of temperature and pressure. As manufacturing processes such as casting involve wide temperature ranges, the material is transforming due to the resulting changes of equilibrium conditions. The transformations caused by a certain thermal history lead to the final microstructure of the system and can be understood with the help of phase diagrams. They show the single-phase regions and the regions, in which two or more phases coexist at equilibrium. The diagrams also contain information about the phase fractions, which can be determined by the application of the lever rule. Invariant reactions take place at defined temperatures and compositions and are of particular interest in material science, as they are related to special characteristics like the formation of lamellar microstructures in the case of eutectic solidification. This kind of solidification happens at a relatively low melting point, the so called eutectic point, which is characterized by the coexistence of one liquid and two solid phases. Another example of an invariant reaction is the peritectic reaction, for which a new solid phase forms out of the liquid phase and the primary solid phase. Corresponding reactions in solid state are the eutectoid and peritectoid reaction.

3.4 Binary systems

If phase transformations happen without a change of composition (allotropic transitions), the equilibrium is defined by equal Gibbs energies of the phases. In the case of varying composition the equilibrium must be determined differently, like it is explained in the following example. For a binary system A-B with two phases α and β the Gibbs energies at a temperature T_1 are drawn in figure 3.10(a) with x as the mole fraction of component B. In this example the entire system has a composition of x_1.

At a first glance one might think, that only phase α is stable, because its Gibbs energy for the overall composition G_1^α is lower than the one of phase β. This would be the case, if both phases were forced to have the composition of the entire system. As both components are completely soluble in both phases, the overall composition can be split up into different compositions of the phases x_1^α

and x_1^β, such that the Gibbs energy of the whole system is lower than G_1^α. Hereby, the composition of the system x_1 has to be split up into the compositions of the phases, such that the mass is conserved. This is the case, if the equation

$$x_1 = f_\alpha x_1^\alpha + f_\beta x_1^\beta \tag{3.20}$$

is fulfilled. Hereby the phase fractions for the system are given by f_α and f_β, with the constraint

$$f_\alpha + f_\beta = 1. \tag{3.21}$$

A visualization of exemplaric phase fractions and compositions can be found in figure 3.9.

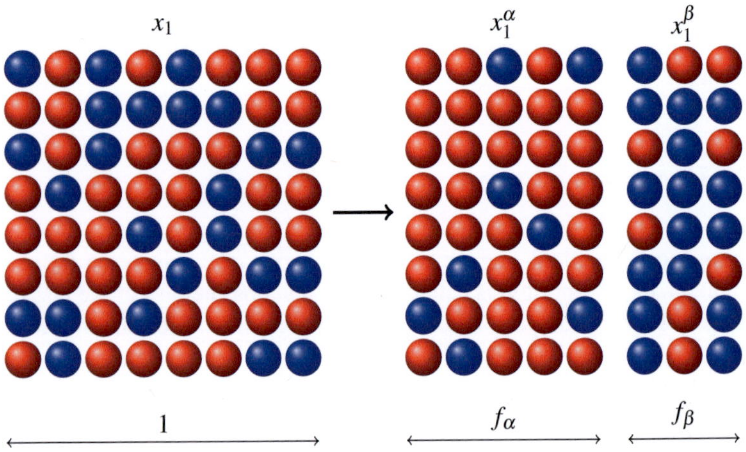

Figure 3.9: A system with the composition x_1 (drawn on the left) splits up into the phases α and β. The separate phases (drawn on the right) have the compositions x_1^α and x_1^β. The phase fractions f_α and f_β are chosen such that the amounts of substance are conserved.

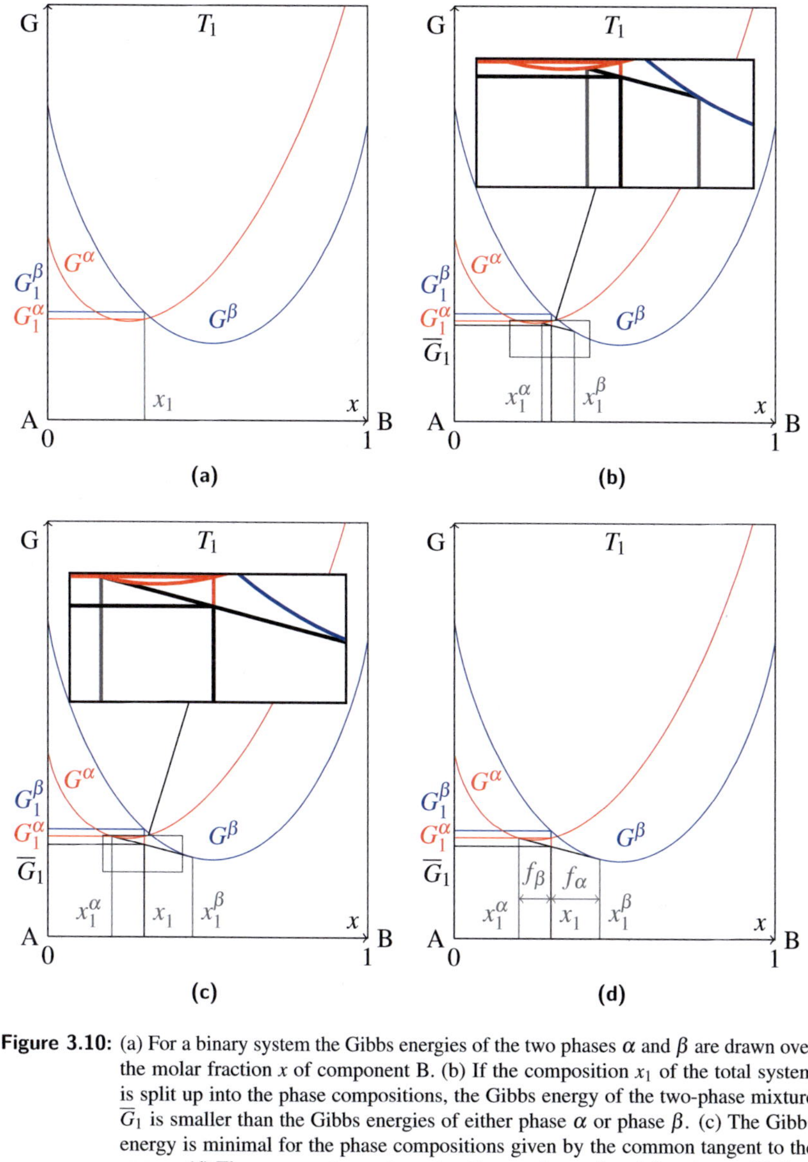

Figure 3.10: (a) For a binary system the Gibbs energies of the two phases α and β are drawn over the molar fraction x of component B. (b) If the composition x_1 of the total system is split up into the phase compositions, the Gibbs energy of the two-phase mixture $\overline{G}_1$ is smaller than the Gibbs energies of either phase α or phase β. (c) The Gibbs energy is minimal for the phase compositions given by the common tangent to the curves. (d) The common tangent construction can be used to define the equilibrium compositions and the phase fractions.

By combining equation 3.20 and 3.21 the lever-rule can be derived, which is given by the equations:

$$f_\alpha = \frac{x_1^\beta - x_1}{x_1^\beta - x_1^\alpha}, \tag{3.22}$$

$$f_\beta = \frac{x_1 - x_1^\alpha}{x_1^\beta - x_1^\alpha}. \tag{3.23}$$

This rule can be used to calculate the phase fractions if the phase compositions are known. To determine the equilibrium phase compositions the Gibbs energies have to be considered. The total Gibbs energy of a system consisting of two phases is given by

$$\overline{G} = f_\alpha G^\alpha(x_1^\alpha) + f_\beta G^\beta(x_1^\beta) = \frac{(x_1^\beta - x_1)G^\alpha(x_1^\alpha) + (x_1 - x_1^\alpha)G^\beta(x_1^\beta)}{x_1^\beta - x_1^\alpha}. \tag{3.24}$$

In the case shown in figure 3.10(b) the two phases have different compositions. The Gibbs energy of the two-phase mixture defined in equation 3.24 can be constructed by drawing a connecting line between $G^\alpha(x_1^\alpha)$ and $G^\beta(x_1^\beta)$. For the composition x_1 the Gibbs energy $\overline{G}$ is given by the point on the connecting line at this composition. As this Gibbs energy is lower than the Gibbs energies of the phases with the composition of the whole system, the system with two phases and different compositions is energetically favorable. Nevertheless this state is not the one with minimal Gibbs energy. If the composition of phase α is lowered and x_1^β is increased, the total Gibbs energy decreases until it reaches a minimum. This minimum is given by the common tangent to the two Gibbs energy curves, like it is shown in figures 3.10(c) and 3.10(d). For these phases compositions the system is in equilibrium. This common tangent construction can be used to construct phase diagrams.

For alloys, phase diagrams show the equilibrium constitutions of the material system and their axis are given by the composition and the temperature. The previously introduced example-system A-B is applied to explain the construction of a simple isomorphous phase-diagram. By the common tangent construction, the two-phase region between x_1^α and x_1^β for the temperature T_1 has already been derived, as shown in figure 3.11(a). The limit of solution of B in phase α is given by x_1^α and analogously x_1^β is the limit of solution for A in phase β. For $x < x_1^\alpha$ and $x > x_1^\beta$ the lowest Gibbs energies are the ones of the single phases and thus there are one-phase regions at the A-rich and B-rich sides. For a higher temperature T_2 again a common tangent can be applied to the curves. The resulting equilibrium compositions have shifted towards higher amounts of component B, as can be seen in figure 3.11(b). At the temperature T_3 the Gibbs energy of phase α is the lowest one for any composition and no common tangent between the phases can be constructed (see figure 3.11(c)). Therefore the α phase is the only stable one for this temperature. If such a construction is performed for all temperatures, a phase-diagram like in figure 3.11(d) can be drawn. It displays the two-phase region of $\alpha + \beta$, which is separated by curves of the equilibrium compositions from the single phase regions. It also includes the tie-lines for T_1 and T_2, which are the connecting lines between the equilibrium compositions. By applying the same principle, more complex diagrams than this simple example can be constructed. However, for multi-component systems the results cannot be represented by two-dimensional graphs as it is possible for binaries.

3.4.1 Gibbs phase rule

The Gibbs phase rule states how many variables are needed to completely describe a system in thermodynamic equilibrium. The state is unambiguously defined, if the number of variables is the same as the degrees of freedom F given by

$$F = 2 + K - P \tag{3.25}$$

for a system of K components and P phases. In materials science the pressure is often assumed to be constant, such that one degree of freedom is already removed and the phase rule reduces to

$$F = 1 + K - P. \tag{3.26}$$

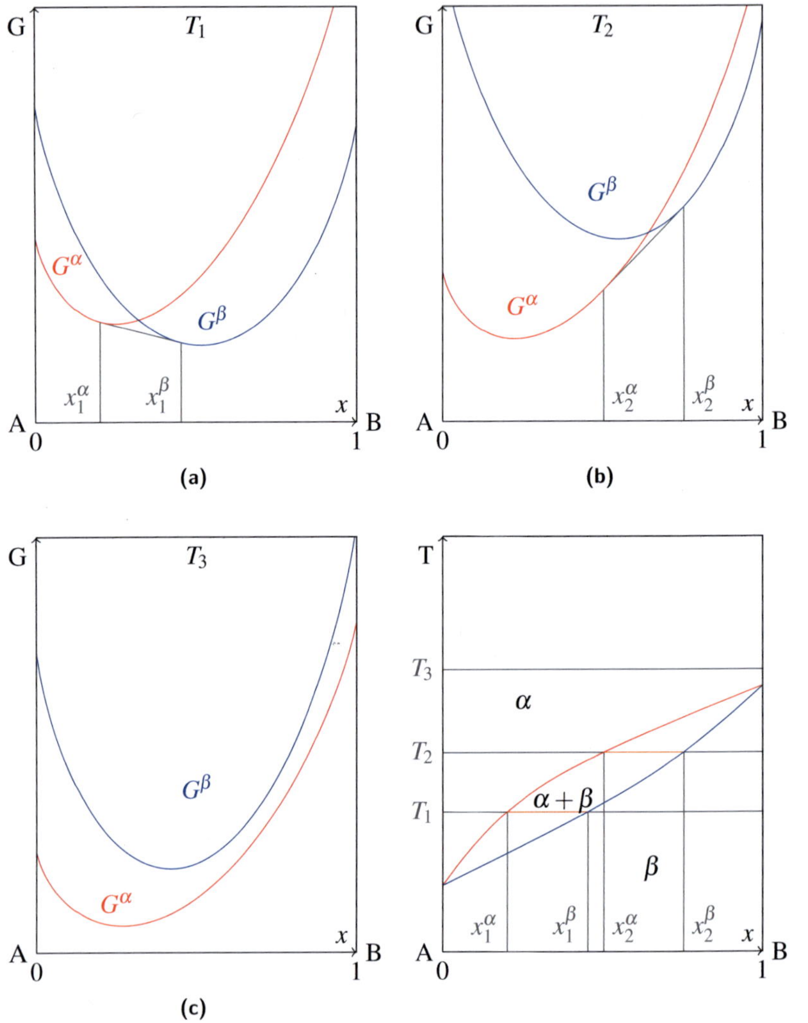

Figure 3.11: Schematic derivation of a phase-diagram. (a) Common tangent of the Gibbs energies for the lowest temperature. (b) At T_2 the equilibrium concentrations have shifted towards a higher amount of B. (c) At the highest temperature only phase α is stable. (d) The constructed phase-diagram with the tie-lines for T_1 and T_2 drawn in orange.

For the binary system in figure 3.11(d), the degrees of freedom are given as

$$F = 3 - P. \tag{3.27}$$

For only one phase in thermodynamic equilibrium the degree of freedom is two. In the phase-diagram this is represented by the two-dimensional one-phase-regions with variable temperature and composition. If phase α and β coexist at a certain temperature, the degree of freedom is only one. This means, that for a variation of temperature their equilibrium compositions have to follow the one-dimensional liquidus or solidus lines (if α is liquid and β is solid).

3.5 Ternary systems

Similarly to a binary phase-diagram, a ternary system can be represented by a three-dimensional diagram, as depicted in figure 3.12. The surfaces of the phase regions are plotted in dependence of temperature and the compositions, which are given by the position in the equilateral Gibbs triangle.

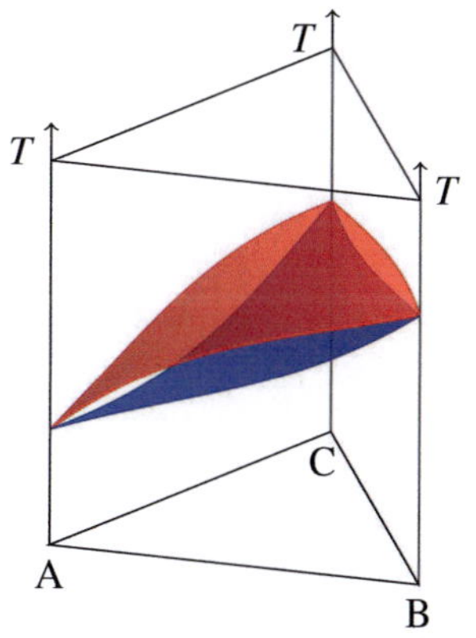

Figure 3.12: The isomorphous ternary system A-B-C is represented by a prism with the equilateral Gibbs triangle as its base. This system is bounded by the three planes of the isomorphous binary systems (see also figure 3.11(d)) and contains the liquidus and solidus surface colored in red and blue, respectively.

A two-dimensional representation of ternary systems is possible by sectional planes through the diagram or by projections of the liquidus surfaces. As an example how to derive a graph of an isothermal section, a simple idealized ternary system with three phases α, β and γ is regarded. For a constant temperature, the Gibbs energies of the three phases are depicted as paraboloids of revolution in figure 3.13(a). They are dependent on the concentration of the components A, B and C, diagrammed by the triangular composition graph on the bottom of figures 3.13(a)- 3.13(d). Equilibria between the different phases are characterized by minimal Gibbs energies. To find these equilibria, the total envelope of the Gibbs energies gets derived step by step in the following. An equilibrium between three phases is given, if the Gibbs energy of their mixture is lower than the Gibbs energies of the pure phases or of mixtures of only two phases. The red triangle in 3.13(b) is spanned between the three contact points of the common tangential plane with the three paraboloids and is a graphical representation of the Gibbs energies of three phase mixtures. In figure 3.13(c) the minimal surfaces spanned between all combinations of two paraboloids of revolution, which are lying on the overall envelope, are colored in blue. These surfaces include the Gibbs energies of the mixtures of two phases. Finally in figure 3.13(d) the remaining surfaces of the paraboloids of revolution, which are part of the total envelope, are shown in orange. A projection of these colored surfaces to the composition triangle at the bottom results in the isothermal section of the ternary phase diagram in figure 3.14. The orange single phase regions are located near the corners of the pure components, the three phase region is in the center of the section and the two phase regions are along the sides of the composition triangle.

3.6 Vertical sections

A binary system A-B consists of all the points, that are uniquely defined by temperature and the composition of one element x_A, for which the equation $x_A = 1 - x_B$ holds. This definition is valid for the condition, that the influence of pressure can be neglected, and in this case it is thus only a two-dimensional problem. Having an additional dimension, a ternary system A-B-C can be graphically represented by a prism like in figure 3.15. This prism is bounded by the planes of the three binary subsystems, which are positioned such that they are orthogonal to any isothermal plane. This orthogonality to any isothermal plane is however not a unique property of binary systems. In principal one can construct an infinite number of vertical sections through a ternary system, as for example in figure 3.15.

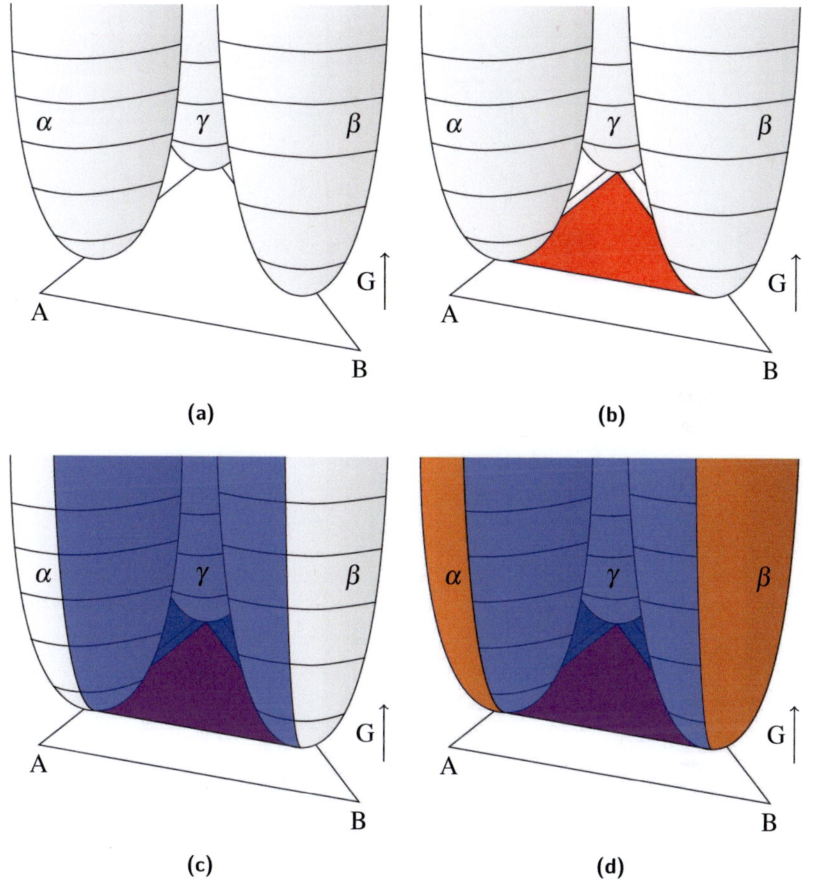

Figure 3.13: Schematic derivation of a ternary isothermal section. (a) Paraboloid Gibbs energies of phases α, β and γ for a fixed temperature. (b) Excerpt of the common tangential plane of all three paraboloids (in red). (c) Minimal surfaces spanned between every pair of paraboloids (in blue). (d) Remaining paraboloid surface (in orange).

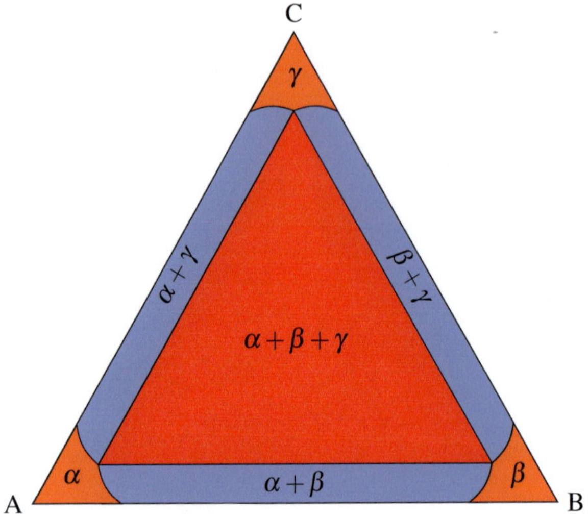

Figure 3.14: Projecting the colored surfaces of figure 3.13(d) onto the concentration simplex results in a graph of an isothermal section.

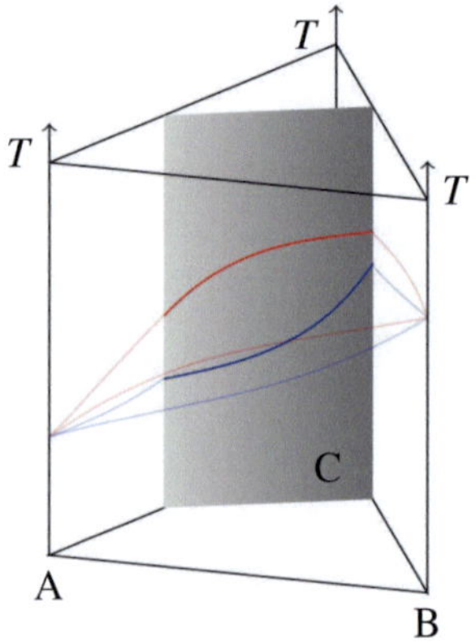

Figure 3.15: A three-dimensional graph of a ternary system A-B-C with a vertical section, which is depicted in grey. The binary planes and the vertical section are orthogonal to the base and to any other plane of constant temperature. The sectional lines of the solidus surface (blue) and liquidus surface (red) are displayed on the vertical section and also shown transparently on the binary planes.

Similarly to a binary system, the points contained in such a subsystem are uniquely defined by the temperature and the composition of one independent component. If one displays the sectional lines of the solidus and liquidus surface on the plane of the subsystem, the resulting graph bears resemblance to a binary phase-diagram. However, a characteristic of a binary system is the property, that all tie lines are located inside the binary plane. For an arbitrary subsystem, the tie lines are in general not located inside the respective sectional plane. For example, the points defined by a section through a liquidus surface are in most cases in equilibrium with points on the solidus surface outside of the sectional plane. This issue is illustrated in figure 3.16. Throughout this thesis, such an arbitrary subsystem is called a pseudobinary system. Non-binary subsystems, for which all the tie lines are located in the sectional plane, have similar properties as binary systems and are therefore called quasibinary systems. Such systems can be found between certain congruently melting binary compounds and pure elements or other such compounds. They can be visualized by phase-diagrams with the binary compound acting like an element. A discussion about the pseudobinary nature of a subsystem of the ternary alloy Al-Cr-Ni and its thermodynamic description for phase-field simulations can be found in section 11.5.

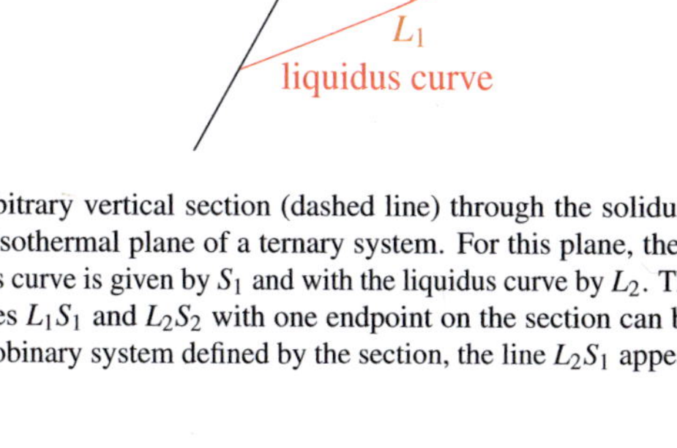

Figure 3.16: An arbitrary vertical section (dashed line) through the solidus and liquidus curves in an isothermal plane of a ternary system. For this plane, the intersection with the solidus curve is given by S_1 and with the liquidus curve by L_2. Therefore, two different tie lines L_1S_1 and L_2S_2 with one endpoint on the section can be constructed. In the pseudobinary system defined by the section, the line L_2S_1 appears to be a tie line.

4 Fundamentals of diffusion

Diffusion plays an important role in many industrial processes. For instance it can be utilized as an alternative to traditional welding for joining metals by the process of diffusion bonding. On the other hand diffusion can also be critical for the intactness of adjoining layers, which are consisting of different materials. To prevent or at least slow down the diffusion from one layer into another, they can be separated by diffusion barriers. In integrated circuits for example a thin layer of Ti-N can act as a diffusion barrier between the conducting lines made of copper and the silicon substrate. This subsection gives an overview of some basic mechanisms of diffusion, which are schematically visualized in figure 4.1. As the focus of this thesis is more on the influence of thermodynamics on solidification, this topic is not addressed in detail.

In general, diffusion denotes the migration of atoms, molecules or charge carriers heading towards a uniform distribution in space. This case, for which the concentration difference is the driving force, is called "downhill diffusion". Consider for example a container with pure water, which gets connected to another one containing a solution of ethanol in water. Then the mechanism of "downhill diffusion" causes the ethanol molecules to "move down" the concentration gradient into the container with pure water and, given enough time, the ethanol concentration becomes statistically equal throughout the whole system. But also the so called "uphill diffusion" is possible, for which the gradients of concentrations are increasing with time. This can for example occur in a monotectic alloy like Fe-Sn, if the liquid phase has an initial composition inside of the spinodal region of the phase-diagram. Then the liquid decomposes into Fe-rich and Sn-rich regions by the process of spinodal decomposition. In all cases however atoms move from higher chemical potentials to lower chemical potentials.

The solid state diffusion in alloys can be classified into interstitial and substitutional diffusion. If an alloy consists of elements, that are of strongly different sizes, the smaller species can diffuse in between the lattice formed by the bigger atoms. This mechanism shown in figure 4.1(a) is called interstitial diffusion. A well-known example for this mechanism is the diffusion of small carbon atoms inside a lattice of iron atoms in the system Fe-C. If the alloying elements are of comparable size and they occupy the same crystal lattice in solid state, the diffusion happens substitutionally. As the name suggests, atoms migrating through the crystal are substituting other atoms, which are located on the same lattice. This exchange process generally incorporates vacancies, which enable the atoms to migrate through the crystal by jumping into them. The substitutional mechanism for example takes place in phases consisting of only the same elements, which is called self-diffusion. Figure 4.1(b) shows a solid phase consisting of pure component A. A way to determine diffusion coefficients for a lattice of pure elements is to measure the tracer diffusion D_i^* of similar microscopic particles, which are radioactive or fluorescent. The tracers have a nearly identical diffusion behavior as the pure elements, but as an advantage they can be detected and distinguished from the matrix atoms. In contrast to self-diffusion, the mechanism of interdiffusion denotes the diffusion of different species of atoms into one another (like in figure 4.1(c)). Such interdiffusion for example occurs in a substitutional manner in the system Cu-Fe, which is addressed in section 15. The flux of component i is given by the version of Fick's first law for multi-components

$$J_i = -\sum_{j=1}^{K} D_{ij} \nabla c_j \tag{4.1}$$

with the concentration $c_j = x_j / V_m$ as amount of substance per unit volume and the interdiffusion coefficient D_{ij}. If the matrix of interdiffusion coefficients contains off-diagonal elements, this cross-coupling can either result in repulsive or in attractive interactions between the different atoms (see also section 14.2.2). The previous equation is based on the assumption, that equilibrium is reached for identical concentrations anywhere in the system. As pointed out before, this is not always the case and a more general formulation relates the fluxes to gradients of the chemical potentials

$$J_i = -\sum_{j=1}^{K} L'_{ij} \nabla \mu_j \tag{4.2}$$

with the phenomenological coefficients given by L'_{ij}. This matrix relating the fluxes to the chemical potential gradients comprises the atomic mobilities M_i of the components, which define the proportionality between the force induced by the gradients of the chemical potentials and the resulting drift velocity of the atoms. The atomic mobilities can be determined from the tracer diffusion by the relation

$$M_i = \frac{D_i^*}{RT}. \tag{4.3}$$

The phenomenological coefficients are employed in the approach of computational kinetics and their definition is given in chapter 5.2. Diffusion coefficients are dependent on composition, temperature, pressure and the structure of the phases. For example the mechanism of substitutional diffusion is enhanced by higher temperatures, because of the increased number of vacancies. In general the diffusion rate for interstitial diffusion is higher than for substitutional diffusion, because the concentration of interstitial atoms is normally low and so they are mostly surrounded by vacancies. For solidification the most important factor is however the state of aggregation. In a liquid phase the diffusion coefficient is several orders of magnitude larger than in a solid phase, as the location and movement of the atoms is not bound to fixed lattice positions (as depicted in figure 4.1(d)).

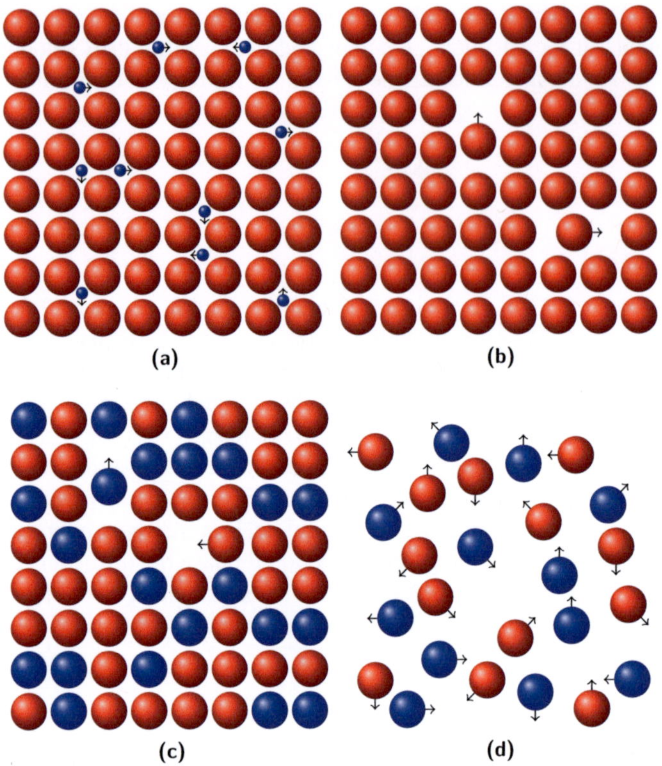

Figure 4.1: Mechanisms of Diffusion: (a) Interstitial diffusion of small atoms in a parent lattice of bigger atoms. (b) Self-diffusion in a solid phase in consequence of vacancy jumps. (c) Interdiffusion of different elements due to the substitutional mechanism. (d) Diffusion in a binary liquid.

5 Computational thermodynamics and kinetics

Phase diagrams are an indispensable tool in material science. Nowadays they are rarely drawn by hand, but are derived by thermodynamic optimization according to the rules, that are explained in section 3.3. For this purpose, the CALPHAD method (CALculation of PHAse Diagrams) was originally made up by Larry Kaufman [2] in 1970. The first software implementation of the CALPHAD approach was done by Lukas in 1977. With the continuous progression of computer technology, the CALPHAD method became a commonly used tool among materials scientists [6]. The algorithms for phase diagram calculation are implemented in commercial software like Thermo-Calc [14], PANDAT [15], FactSage [16] or MTDATA [17] and also in open source software like OpenCalphad [18]. But none of the programs can create a phase-diagram, if it lacks a suitable thermodynamic dataset as input. The pool of available thermodynamic data is constantly increasing by the contributions of researchers from all over the world. To improve the communication of the latest assessments and to promote collaborations, a CALPHAD journal [19] and an annual conference were organized.

For commonly used alloys the databases are well-tried and reliable, as they have been gathered from various experiments and ab initio calculations. However, the availability of thermodynamic information decreases with every additional alloying element. The strong information content of the datasets not only enables the calculation of exact phase-diagrams, but also increases the quantitativeness of computer simulations coupled to thermodynamic databases, as addressed in section III. Today the field of computational thermodynamics covers the assessment, the storage and the application of thermodynamic data, e.g. in the branch of computational kinetics. This method can be used for the simulation of diffusional processes, by combining thermodynamic and mobility data.

5.1 CALPHAD method

The particular strength of the CALPHAD method lies in the fact, that the thermodynamic properties of a system for a state, for which no experimental data is available, can be extrapolated from present data of other states. Dinsdale started to record datasets for pure elements (unary systems) [20], which are now included in the datasets for binaries. The data for binary systems can be used to derive ternary ones and these can be used again to build up databases for multi-component systems of higher order. Thereby the data quality of the basic "building blocks" – the unary and binary systems – is essential for the accuracy of the higher-order systems [21]. If new data is added to any of the subsystems, it can be directly used to update the higher order systems. But not only the elements from the periodic table can be treated in the CALPHAD approach, also combinations of elements like H_2O or ions like Fe^{3+} can be defined as the constituents of the phases.

To calculate phase diagrams based on thermodynamic functions requires to find the equilibrium of a material system for certain conditions, characterized by a minimum of an appropriate thermodynamic potential. The potential used in the CALPHAD method is the Gibbs free energy as its characteristic state variables: temperature T, pressure p and composition $\mathbf{x}$ are known for most thermodynamic measurements. Hereby and in the following, $\mathbf{x} = \{x_i\}_{i=1}^{K}$ stands for a vector including the mole fractions $x_i = N_i/N$ of a system with K components. Since there is only a small influence of pressure on the Gibbs energy of solid and liquid phases, it is usually neglected, just like in the following. For the purpose of phase diagram calculation of multi-component systems, the models for the Gibbs energies introduced in 3.2 are often too simple. The descriptions need to reproduce various data obtained from experiments and also from ab initio calculations in a concise way. Therefore sophisticated formalisms have been developed, which describe the Gibbs energies G_m^{α} for phases α over the whole composition range and in the temperature regime of technical interest. The subscript m hereby indicates, that the energy is referred to one mole of a substance. The same applies for the further thermodynamic variables, which can be derived from the molar Gibbs energies as described hereinafter.

5.1.1 Data assessment

Any thermodynamic dataset needs to be based on well-founded material data, to reproduce the correct phase-diagrams. Therefore the big variety of available data from experiments or first-principle calculations has to be critically checked for compatibility before it is included in the datasets. A source of data is given by crystallographic information about the structure, the sublattices and defects of a material. Also data, that can be directly represented by phase diagrams, is taken into account. This consists of the temperatures, at which phase transformations happen, the microscopical informations about the distribution of phases, and furthermore the techniques of microprobe, X-ray and neutron diffraction. An important role plays the assessment of thermochemical properties of the material. The technique of differential scanning calorimetry can be used to get the heat content, the heat capacity or the enthalpy of formation. Chemical potentials and activities can for example be derived from measuring the electromagnetic field of galvanic cells. This data can be used to determine the coefficients of the chosen Gibbs energy formulation, like the one given in equation 3.9. Further physical data is methodically collected to determine the magnetic parameters for the Gibbs energy model explained in section 5.1.8. Beside experimental results, first-principle calculations are also used to extend the pool of thermodynamic data. This is especially beneficial to obtain informations about states, for which experiments would be extremely expensive or impossible. For example the density-functional theory can be used to obtain thermodynamic quantities at absolute zero. Aside from proprietary file formats, thermodynamic datasets are usually available as so-called TDB (Thermodynamic DataBase) files, which is a plain-text format. In these files, the thermodynamic information is stored as Gibbs energy functions, as described in the following.

5.1.2 Gibbs energy formulation

A key aspect of the CALPHAD method is to exactly describe the measured thermodynamic quantities by a suited model of the Gibbs energies. Within the scope of the CALPHAD approach, the total Gibbs energy G_m^α of a phase α is modeled phenomenologically as a sum of up to four parts

$$G_m^\alpha(T,\mathbf{x}) = {}^{ref}G_m^\alpha(T,\mathbf{x}) + {}^{id}G_m^\alpha(T,\mathbf{x}) + {}^{ex}G_m^\alpha(T,\mathbf{x}) + {}^{phys}G_m^\alpha(T,\mathbf{x}). \tag{5.1}$$

Hereby the reference part ${}^{ref}G_m^\alpha$ is given as a linear interpolation over the composition between the Gibbs energies of the end-members and corresponds to the model of mechanical mixing in equation 3.13. In the CALPHAD method end-members denote pure elements or compounds with a fixed composition and their Gibbs energies can be expressed with temperature dependent formulations like in equation 3.9 (as mentioned before, the pressure dependence is not taken into account in this thesis). If a phase consists of several sublattices, an end-member is given, if each sublattice consists of only one constituent [22]. ${}^{id}G_m^\alpha$ is the ideal mixing part, which takes the entropy of mixing of the components into account, like in equation 3.17. The excess part ${}^{ex}G_m^\alpha$ is the remainder of the subtraction of all other parts from the total Gibbs energy. It represents all contributions, for which the use of a suitable physical model would lead to an inappropriate complexity. If there is on the other hand a concise description for a physical phenomenon like the ferromagnetic transition, it can be appended as an additional part ${}^{phys}G_m^\alpha$ instead of representing it by an excess part of higher order. In the next paragraphs, the specific formulations of the individual terms from equation 5.1 are given.

5.1.3 Compound-energy formalism

Most of the models for the Gibbs energy are a subset of the general compound-energy formalism (CEF), which is explained in a paper from Mats Hillert [22]. This model was contrived by Hillert and Staffanson [23] and extended by Sundman and Ågren [24] and can be applied to many different materials like intermetallic phases, interstitial solutions or carbides. Sometimes the CEF is also called the sublattice model, as it can handle an arbitrary number of components, which can be located on an arbitrary number of sublattices. The sublattice description is based on the actual crystallographic information of the material systems. As the materials can have a lot of different lattices, in most cases several real lattices are represented by only one sublattice in the CALPHAD context. The CEF provides the following expressions for the individual parts of the Gibbs energy:

$$ {}^{ref}G_m^\alpha(T,Y) = \sum_{I_0} P_{I_0}(Y)\, {}^{\circ}G_{I_0}^\alpha(T) \tag{5.2} $$

$$ {}^{id}G_m^\alpha(T,Y) = RT\sum_{s=1}^{n} a_s \sum_{i=1}^{n_s} y_i^s \ln\left(y_i^s\right) \tag{5.3} $$

$$ {}^{ex}G_m^\alpha(T,Y) = \sum_{I_1} P_{I_1}(Y) L_{I_1}^\alpha(T) + \sum_{I_2} P_{I_2}(Y) L_{I_2}^\alpha(T) + \ldots \tag{5.4} $$

where R is the gas constant. This formalism can treat a phase α having n different sublattices with every sublattice s having a_s sites. These sites can be occupied by the n_s different constituents being present in this sublattice. So the composition $\mathbf{x}$, given as mole fractions, splits up into site fractions y_i^s over the different sublattices. Y is a matrix containing the site fractions of all components in all sublattices and was introduced by Sundman and Ågren [24] together with the concept of constituent arrays I_k. This concept denotes the distribution of the constituents in the sublattices of a phase for the application in computer calculations. The subscript k hereby defines the order of the array, which corresponds to the number of independent site fractions. For a given constituent array I_k the function $P_{I_k}(Y)$ returns the product of all nonzero site fractions in Y. Furthermore ${}^\circ G_{I_0}^\alpha(T)$ is the Gibbs energy of formation for an end-member defined by the constituent array I_0 and $L_{I_k}^\alpha(T)$ is the interaction parameter for a constituent array I_k. These parameters describe the influence of interactions in higher order systems and are designed such that they can be determined from available data of lower order systems. This approach was proposed by Redlich and Kister [25] and the interaction parameters are therefore also referred to as Redlich-Kister parameters. The temperature dependence of the reference part and the excess part is modeled solely through the Gibbs energies of formation and the interaction parameters, which gets explained in section 5.1.6.

5.1.4 Simplified formulation for the binary case

The general framework of the compound-energy formalism enables the description of Gibbs energies for a wide range of material systems. For the application within the phase-field method in the case of a binary system, it is however helpful to start from a simplified formulation. The following equations are valid for phases α of a binary system A-B, which are modeled without sublattices or additional physical contribution parts. As the mole fraction of component B can be expressed as $1-x$, the Gibbs energies of these phases only depend on temperature and on the mole fraction x of the first component A. For that case, the CEF reduces to this regular-solution type model:

$$^{ref}G_m^\alpha(T,x) = x \cdot {}^\circ G_A^\alpha(T) + (1-x) \cdot {}^\circ G_B^\alpha(T) \tag{5.5}$$

$$^{id}G_m^\alpha(T,x) = RT\,(x\ln(x) + (1-x)\ln(1-x)) \tag{5.6}$$

$$^{ex}G_m^\alpha(T,x) = x(1-x)\sum_{\nu=0}^{k} {}^\nu L_{A,B}^\alpha(T) \cdot (x-(1-x))^\nu \tag{5.7}$$

Here ${}^\circ G_A^\alpha(T)$ and ${}^\circ G_B^\alpha(T)$ are the Gibbs energies of the pure components in phase α and ${}^\nu L_{A,B}^\alpha(T)$ are the binary interaction parameters of order ν for this phase. Furthermore the derivative of the total Gibbs energy with respect to x is given by the sum

$$\frac{\partial G_m^\alpha(T,x)}{\partial x} = \frac{\partial\, ^{ref}G_m^\alpha(T,x)}{\partial x} + \frac{\partial\, ^{id}G_m^\alpha(T,x)}{\partial x} + \frac{\partial\, ^{ex}G_m^\alpha(T,x)}{\partial x} \tag{5.8}$$

including the derivatives of the three individual contributions:

$$\frac{\partial\, ^{ref}G_m^\alpha(T,x)}{\partial x} = {}^\circ G_A^\alpha(T) - {}^\circ G_B^\alpha(T) \tag{5.9}$$

$$\frac{\partial\, ^{id}G_m^\alpha(T,x)}{\partial x} = RT\ln\left(\frac{x}{1-x}\right) \tag{5.10}$$

$$\frac{\partial\,{}^{ex}G_m^\alpha(T,x)}{\partial x} = 2(x-x^2)\sum_{\nu=0}^{k} {}^{\nu}L_{A,B}^{\alpha}(T)\cdot \nu(2x-1)^{\nu-1} - \sum_{\nu=0}^{k} {}^{\nu}L_{A,B}^{\alpha}(T)(2x-1)^{\nu+1}. \tag{5.11}$$

5.1.5 Ternary contribution of the excess part

Following the approach of Muggianu et al. [26], the Redlich-Kister formulations can be extended for higher-order systems. In case of ternary systems, the excess part in equation 5.1 is modeled by an additional term

$$^{ex,\,te.}G_m^\alpha(T,\mathbf{x}) = \sum_{i=1}^{n-2}\sum_{j=i+1}^{n-1}\sum_{k=j+1}^{n} x_i x_j x_k \left(\nu_i\cdot {}^{i}L_{ijk}^{\alpha} + \nu_j\cdot {}^{j}L_{ijk}^{\alpha} + \nu_k\cdot {}^{k}L_{ijk}^{\alpha}\right), \tag{5.12}$$

including the ternary interaction parameters ${}^{i}L_{ijk}^{\alpha}$, ${}^{j}L_{ijk}^{\alpha}$ and ${}^{k}L_{ijk}^{\alpha}$. These parameters represent the change in Gibbs energy due to the interaction of three components. To take care of the ternary interactions for systems with more than three components, the ν_i fractions introduced by Hillert [27] can be used:

$$\nu_i = x_i + \frac{1-x_i-x_j-x_k}{3} \tag{5.13}$$

$$\nu_j = x_j + \frac{1-x_i-x_j-x_k}{3} \tag{5.14}$$

$$\nu_k = x_k + \frac{1-x_i-x_j-x_k}{3}. \tag{5.15}$$

In the case of ternary systems these fractions reduce to the molar fractions.

5.1.6 Parameters dependent on temperature

The Gibbs energies of formation in equation 5.2 depend on temperature and are modeled with power series like in equation 3.9. These parameters are adjusted to fit the data from experiments as described in section 5.1.1. To avoid a large number of coefficients, they are chosen as piecewise functions of temperature, which have to be at least two times continuously differentiable at the breakpoints. For stable end-members, the temperature dependence of the molar Gibbs energy is usually described relative to the sum over the enthalpies of all constituents of the end-member H_i^{SER}, weighted with the respective stoichiometry factor b_i. The enthalpies are typically given for the standard element reference state, abbreviated SER, which is characterized by a temperature of 298.15 K and a pressure of 1 bar. Inside a certain temperature range between T_n and T_{n+1} such a power series often has the form of

$$^{\circ}G_{I_0}^{\alpha}(T) - \sum_i b_i H_i^{SER} = d_0 + d_1 T^{-1} + d_2 T \ln(T) + \sum_{n=1} e_n T^n \tag{5.16}$$

with the coefficients d_n and e_n. For unstable end-members, the description of the Gibbs energies of formation is mostly given in relation to the Gibbs energies of the so called reference phases, which are the stable ones at the chosen reference state for the case of a unary system.

The Redlich-Kister parameters $L_{I_k}^{\alpha}$ used in the excess part of the CEF are often chosen to be linear in temperature:

$$L_{I_k}^{\alpha}(T) = d_0 + d_1 T. \tag{5.17}$$

This is because according to Lukas, Fries and Sundman [12, p.109] heat capacity data needed to model a temperature dependence of higher order is not available in many cases. The temperature dependence of the ternary interaction parameters in equation 5.12 is modeled in a similar way.

5.1.7 Modeling of stoichiometric phases in CALPHAD

In the CALPHAD method, the Gibbs energy of a stoichiometric phase β can be defined only for the particular stoichiometric composition x^β as a function of temperature like in equation 3.9. This modeling approach is depicted in figure 5.1(a) for a certain temperature. In a phase diagram derived from such a formulation, the phase appears as a perfect vertical line. An alternative way is to model a stoichiometric phase as a composition dependent function with a strong curvature, as it is done for the phase shown in figure 5.1(b). If the minimum of the curve is set at the stoichiometric composition, the composition of the stoichiometric phase in equilibrium with another phase only varies by a small difference δx^β from x^β. In the same way the equilibrium composition of the other phase varies slightly by δx^α. The phase is thus not a perfect line compound, as it is stable in a narrow range around the stoichiometric composition.

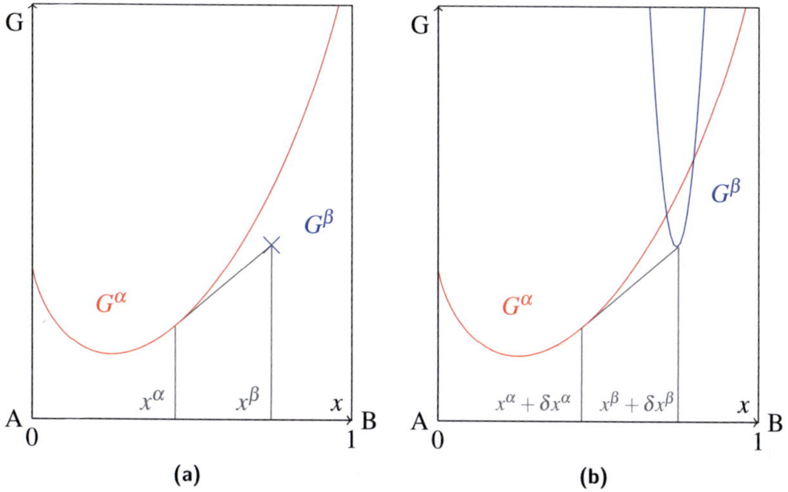

Figure 5.1: Common tangent constructions between a phase α and two variants of a stoichiometric phase β. The Gibbs energy of the stoichiometric phase G^β is only defined at the composition x^β in subfigure (a), whereas it is modeled as a function with high curvature in subfigure (b).

These two different formulations are for example applied to model the silicon rich diamond phase. In an earlier publication about binary systems of transition metals from Larry Kaufman [28], the diamond phase is described as a perfectly stoichiometric crystal containing only silicon. In a later assessment for the system Al-Mg-Si from Feufel et al. [29] (see also section 14.1), the phase is modeled with the Redlich-Kister-Muggianu formulation [25, 26] and has a small solubility of aluminum. A modeling of the diamond phase for the system Al-Si-C can be found in Gröbner et al. [30], in which fictive values are assigned to the Redlich-Kister interaction coefficients to achieve negligible solubility.

5.1.8 Magnetic contribution in the CALPHAD method

Some elements like iron, cobalt or nickel undergo magnetic transitions at some critical temperatures. This transition is accompanied with a discontinuity of the heat capacity and also has a contribution to the Gibbs energy. To take this ferromagnetic or antiferromagnetic behavior into account, an additional term can be included in the Gibbs energy formulation given by equation 5.1. The model, which is used for the magnetic contribution today, was formulated by Hillert and Jarl [31] and reads

$$^{mag}G_m^{\alpha}(T,\mathbf{x}) = RT\ln(\beta+1)\cdot g(\tau) \tag{5.18}$$

with β as the average value of the magnetic moment and the variable τ as the temperature divided by the critical temperature

$$\tau = \frac{T}{T_C}. \tag{5.19}$$

The critical temperature T_C is given by the Curie temperature for ferromagnetic materials and by the Néel temperature for antiferromagnetic materials. The composition dependence of the critical temperature and β are modeled with Redlich-Kister formulations, similar to the parameters used for the other contributions to the Gibbs energy:

$$\beta = x_A\beta_A + x_B\beta_B + x_Ax_B \sum_{\nu=0} \beta_{AB}^{\nu}(x_A - x_B)^{\nu} \tag{5.20}$$

$$T_C = x_AT_{C,A} + x_BT_{C,B} + x_Ax_B \sum_{\nu=0} T_{C,AB}^{\nu}(x_A - x_B)^{\nu} \tag{5.21}$$

with β_A, β_B, $T_{C,A}$ and $T_{C,B}$ as the parameters for the pure components and β_{AB}^{ν} and $T_{C,AB}^{\nu}$ as the interaction parameters of degree ν for the magnetic moment and the critical temperature, respectively. In dependence whether the temperature is below or above the critical temperature, the function $g(\tau)$ is given as

$$g(\tau) = \begin{cases} 1 - \left[\dfrac{79\tau^{-1}}{140p} + \dfrac{474}{497}\dfrac{1-p}{p}\left(\dfrac{\tau^3}{6} + \dfrac{\tau^9}{135} + \dfrac{\tau^{15}}{600}\right)\right]\dfrac{1}{D} & \text{if } \tau < 1 \\ -\left[\dfrac{\tau^{-5}}{10} + \dfrac{\tau^{-15}}{315} + \dfrac{\tau^{-25}}{1500}\right]\dfrac{1}{D} & \text{if } \tau \geqslant 1 \end{cases} \tag{5.22}$$

together with

$$D = \frac{518}{1125} + \frac{11692}{15975}\left(\frac{1}{p} - 1\right). \tag{5.23}$$

Here, the influence of the crystallographic structure enters the equation by the empirical constant p, which is defined as 0.4 for BCC phases and as 0.28 for FCC phases. The magnetic contribution can be considered the most common additional term, although there exists a variety of models to account for specific physical phenomena.

5.1.9 Further models

Over the years more specific formulations have been developed and are used, when it comes to describe special material properties. A comprehensive documentation of these models can be found in the book "Computational thermodynamics" [12]. For example, charged particles like cations and anions can be taken into account as the constituents of the ionic liquid model. This model is a modification of the sublattice model and maintains electroneutrality with a variable number of sites. Another example is the associated solution model, which includes fictitious constituents for modeling short range order. In common datasets, these models are rarely applied and they are especially not relevant for the material systems, which are treated in this thesis. Therefore, they are not examined in more detail hereinafter.

5.2 Computational Kinetics

For the calculation of phase-diagrams, only equilibrium between phases matters and no temporal evolution has to be considered. In contrast, the change over time is treated in computational kinetics. In this approach the thermodynamic information from the CALPHAD method is combined with mobility data for the determination of the diffusion coefficients. These mobility parameters have to be fitted in an assessment process similar to the one for the thermodynamic parameters to give a close match to experimental results.

Computer programs can read the thermodynamic and kinetic datasets and use it to solve models of diffusion driven reactions like carburizing, homogenization or coarsening. A popular software for the calculation of diffusivities in alloy systems and the numerical solution of diffusion equations is DICTRA (DIffusion Controlled TRAnsformations) [14, 32]. This piece of software is designed as a module of the Thermo-Calc program, such that it can directly receive the thermodynamic factors needed to calculate the diffusion coefficients. Because the kinetic parameters are expressed as Redlich-Kister formulations and power series, the routines for the calculation of the thermodynamic quantities can be reused for the purpose of computational kinetics. With programs like DICTRA, the simulations of diffusion driven phenomenons can only be performed for simple geometries like spheres or infinitely long plates or cylinders. This limitation opens up a possible field of application for phase-field simulations to handle complex microstructures. The basic simulation results from DICTRA could also be used to compare with the outcomes from the phase-field solver to check their validity.

Kinetic databases are available for the solid phases of common alloy systems with multiple components. For liquids, the measurement of diffusivities is a difficult task and accurate mobility data is rarely available for them. Therefore the diffusion coefficients assigned to liquid phases in the databases are often given by the commonly used value of 10^{-9} m/s^2. The databases contain coefficients for the mobilities, because in multi-component systems they can be stored more compactly than the diffusion coefficients. According to Borgenstam et. al [32] it requires to store $(K-1)^2$ interdiffusion coefficients but only K mobilities in a system of K components. All the kinetic coefficients in this chapter are defined for a specific phase, but for the sake of clarity no phase index is assigned to them. The independent mobility coefficients M_A of an element A are expressed as

$$M_A = \frac{M_A^\circ}{RT} \exp\left(\frac{-Q_A}{RT}\right) = \frac{1}{RT} \exp\left(\frac{RT \ln(M_A^\circ)}{RT}\right) \exp\left(\frac{-Q_A}{RT}\right) \tag{5.24}$$

with the frequency factor M_A° and the activation enthalpy Q_A. This formulation suggested by Andersson and Ågren [33] does not include the influence of the ferromagnetic transition, which can be treated by an additional factor. The composition dependence of the mobility parameters Φ_A is modeled as a Redlich-Kister formulation,

$$\Phi_A = \sum_{i=1}^{K} x_i \Phi_A^i + \sum_{i=1}^{K} \sum_{j=2,j>i}^{K} x_i x_j \left(\sum_{v=0} {}^{v}\Phi_A^{i,j} (x_i - x_j)^v \right) \tag{5.25}$$

with the parameters for the pure components Φ_A^i and the interaction parameters ${}^{v}\Phi_A^{i,j}$. If the ferromagnetic transition is not considered, the frequency factor and the activation enthalpy can be expressed by one single parameter, which is given by the expression

$$\Phi_A = -Q_A + RT \ln(M_A^\circ). \tag{5.26}$$

In general $\ln(M_A^\circ)$ and Q_A can be given by separate parameters. Analogous to the modeling of the Gibbs energies in the CALPHAD approach, these parameters are stored in the datasets as power series of temperature.

From the mobilities one can obtain the diagonal matrix of the so-called phenomenological parameters L_{ij}. The entries can be calculated as the mobility of a component multiplied with its concentration

$$L_{kk} = c_k M_k. \tag{5.27}$$

This matrix could be used to calculate the fluxes in a lattice-fixed frame of reference. To get the fluxes in reference to the volume or the number of particles, the matrix L' has to be calculated from L. For the case of substitutional elements with equal molar volumes V_m, the following equation holds:

$$L'_{ki} = \sum_{j=1}^{K} \left(\delta_{jk} - c_k V_j\right) L_{ji} = \sum_{j=1}^{K} \left(\delta_{jk} - x_k \frac{V_j}{V_m}\right) L_{ji} \tag{5.28}$$

with V_j as the partial molar volume of component j and the Kronecker symbol δ_{jk} being 1 for $j = k$ and 0 in all other cases. Finally this matrix can be used to derive the interdiffusivities

$$D_{kj} = \sum_{i=1}^{K} L'_{ki} \frac{\partial \mu_i}{\partial c_j}. \tag{5.29}$$

For the calculation of the diffusion coefficients, the thermodynamic factor $\frac{\partial \mu_i}{\partial c_j}$ is required, which can be derived from the CALPHAD method. The thermodynamic factor is given as the partial derivative of the chemical potential with respective to concentration. An example for the application of diffusivities from computational kinetics is given in section 15.

6 Analytical models for solidification

For being able to predict and adjust solidification processes, various analytical models were developed throughout the 20th century and have been checked for validity. The established theories stood the proof to produce reasonable results for the specific types of solidification they are designed for. However, the more the idealizations, which are assumed in the theories, differ from reality, the less accurate are the resulting predictions. The coming of simulation techniques like the phase-field method enabled the treatment of more complicated solidification patterns, than it is possible with analytical models. With a continuing increase of computer power, these simulation methods will be able to cope with higher and higher levels of complexity. But before one approaches large setups by high performance computing, it has to be ensured, that the simulation methods yield the same results as the analytical solutions, if their assumptions are fulfilled. In the scope of this thesis, simulation results are compared with two analytical models, the Mullins Sekerka and the Lipton-Glicksman-Kurz theory. The theoretical foundations of these models are shortly explained in the following.

6.1 Mullins Sekerka theory

Starting from a protrusion, a planar growth front can transform into different morphologies like cellular, dendritic or fractal patterns. The stability of the initially flat interface is determined through the feedback of the concentration and temperature fields ahead of the protrusion. The conditions for a transition from planar to cellular growth can be analyzed with the well-established theory of Mullins and Sekerka [8] (henceforth referred to as MS). Beside the constitutional effects, this theory also incorporates the influence of surface energy. On the one hand a planar front is preferred against a curved interface, which is energetically disadvantageous. On the other hand the undercooled melt provides better growth conditions for protrusions, what causes their amplification. With the Gibbs-Thomson and the Stefan condition as the boundary conditions at the interface, this theory provides an equation for the rate of change of the protrusion's amplitude for different perturbation wavelengths.

The MS theory assumes a flat solidification front growing at steady-state into the z-direction with the velocity v, as it is depicted in figure 6.1(a). The planar interface is characterized by the gradient G_c of the concentration field in the liquid at the interface and by the thermal gradients G and G' at the interface in the liquid and solid, respectively. As a simplification of the analysis from MS it is assumed, that the gradients of temperature only have a minor influence on the stability of the interface and solely consider the gradients of the composition field. A plot of the concentration profile of the front at steady-state can be found in figure 6.1(b). The MS theory gives a prediction about the stability of this front for the case, that it gets slightly disturbed. Hereby, the perturbation assumed in the theory has a sinusoidal shape of frequency ω and wavelength $\lambda = 1/\omega$ and is shown in figure 6.1(c). The interface position of the perturbed front in z-direction can be expressed by the following function of time t and the x-coordinate

$$z(x,t) = \delta(t)\sin(\omega x). \tag{6.1}$$

It is assumed, that the sinusoidal ripples are initialized with an infinitesimal amplitude δ, which afterwards varies with time. The perturbation of the solidification front is of a kind, that the steady-state profiles of the concentration field in the liquid ahead of the front are shifted together with the solid-liquid interface. This ensures, that for any cross section through the rippled interface the profiles have the same shape as for the flat front. In figure 6.1(b) it can be seen, that the differences in concentration between the profiles decay with increasing distance from the interface (having a concentration c_0 in the liquid), as they all converge against the concentration c_∞ in the liquid far away from the front. For this reason Mullins and Sekerka approximate the concentration in the liquid ahead of the front by a linear super-position of the profile of the flat front at steady-state $c_{planar}(x,z)$ and a sine-wave with an amplitude C_1 decaying in z-direction:

$$c(x,z) = c_{planar}(x,z) + C_1 \sin(\omega x) e^{-k_\omega z}. \tag{6.2}$$

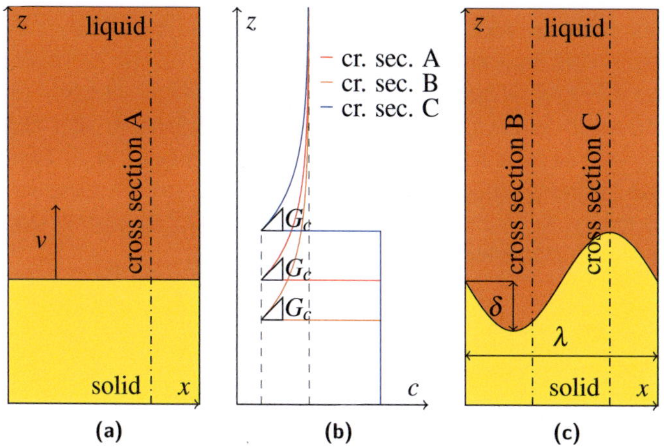

Figure 6.1: (a) Flat solidification front growing at steady state (b) Concentration profiles for different cross sections (c) Sinusoidal perturbation of the steady state front

For a given frequency ω the corresponding decay rate is k_ω. If different wavelengths are applied for the perturbation, the front either returns back to the planar growth mode or reacts with an amplification of the waves. A prediction for these behaviors is given by Mullins and Sekerka with the equation for the rate of change of the amplitude, which can be calculated as

$$\frac{\dot{\delta}}{\delta} = \left[k_\omega - \frac{vp}{D^{liq}}\right]\left(-\frac{D^{liq}G}{mc_0p} - \frac{D^{liq}\Gamma\omega^2}{mc_0p} + vg(\omega)\right) \tag{6.3}$$

with the inter-diffusion coefficient of the liquid phase D^{liq} and the Gibbs-Thomson coefficient

$$\Gamma = \frac{m\sigma_{eff}\frac{\partial c^{liq}(\mu,T)}{\partial \mu}}{\Delta c}. \tag{6.4}$$

The properties of the phase-diagram enter the previous equations with the liquidus slope m, the partition coefficient $k = 1 - p$, the difference of the equilibrium concentrations of the solid and the liquid Δc and the derivative of the liquid concentration with respect to the chemical potential. In the previous formula σ_{eff} is the effective stiffness of the solid-liquid interface, which is equivalent to the stiffness σ in the case of isotropic surface energies. For phase-field simulations with an anisotropy of the surface energy defined like in equation 7.18, the effective stiffness is dependent on the strength of the surface energy anisotropy $\delta_{\alpha\beta}$ according to the formula $\sigma_{eff} = \sigma\left(1 - 15\delta_{\alpha\beta}\right)$. The decay rate of the frequency can be calculated as

$$k_\omega = \frac{v}{2D^{liq}} + \sqrt{\left(\frac{v}{2D^{liq}}\right)^2 + \omega^2} \tag{6.5}$$

and $g(\omega)$ is given by

$$g(\omega) = 1 - \frac{2k}{\sqrt{1 + \left(\frac{2D^{liq}\omega}{v}\right)^2} + 2k - 1}. \tag{6.6}$$

For the isothermal case ($G = 0$, $G' = 0$), equation 6.3 reduces to

$$\frac{\dot{\delta}}{\delta} = \left[k_\omega - \frac{vp}{D^{liq}}\right]\left(-\frac{D^{liq}\Gamma\omega^2}{mc_0 p} + vg(\omega)\right). \tag{6.7}$$

The stability of a front is given by equation 6.3 for the case $\dot{\delta} < 0$, which denotes, that the amplitude of a given perturbation decreases with time and vanishes. On the other hand, a protrusion amplifies and destabilizes the planar front for positive values of $\dot{\delta}$. With the condition $\dot{\delta} = 0$, a critical wavelength λ_0 can be determined, which marks the limit of stability. As phase-field models usually include all the features assumed in the MS theory, simulations should be able to reproduce the analytical results. An application of this stability criterion can be found in the study of cellular growth in section 14.2.

6.2 LGK theory

The growth of thermal or solutal dendrites is strongly dependent on the processes in the tip region. By the rejection of heat and solute, diffusion fields build up around the dendrite tips, which have a major influence on the growth rates. An analytical solution for this type of solidification is given by the Lipton-Glicksman-Kurz (LGK) theory [9] relating the supersaturation to the radius r and velocity v of the tip. A principal feature of the analytical theory is the choice of the function to approximate the shape of the growing dendrite. For the approximation of the tip geometry two strategies are proposed in the book of Kurz and Fisher [34], both assuming an isothermal needle crystal without sidearms. The first approach (referred to as $\mathrm{LGK}_{3\mathrm{D}}$) approximates the 3D shape with a paraboloid of revolution described by the Ivantsov integral [35]

$$I(\mathrm{Pe}) = \mathrm{Pe}\,\mathrm{E}_1(\mathrm{Pe})e^{\mathrm{Pe}}, \tag{6.8}$$

with Pe being the Péclet number $\frac{vr}{2D^{liq}}$. A second expression is provided by Horvay and Cahn [36] for a parabolic plate, where the integral is derived as

$$I(\mathrm{Pe}) = \sqrt{\pi\,\mathrm{Pe}}\,\mathrm{erfc}(\sqrt{\mathrm{Pe}})e^{\mathrm{Pe}}. \tag{6.9}$$

In the following, this two-dimensional solution is denoted by $\mathrm{LGK}_{2\mathrm{D}}$. The Ivantsov solution makes use of the exponential integral function E_1 and the Horvay-Cahn solution of the complementary error function erfc.

By equating one of these shape functions with the undercooling, a relation between the undercooling and the radius and velocity can be obtained. This equation on its own predicts an infinite number of possible combinations of tip radii and velocities for a given undercooling. For example a sharp tip could grow at a high velocity or a blunt tip at a low one. A lower limit for the tip radius is given for the case, when the supersaturation and the curvature balance out. To define the pair of v and r, at which a specific system operates, an additional criterion is needed. For this purpose Langer and Müller-Krumbhaar [37] proposed a criterion

$$\sigma^* = \left(\frac{\lambda_0}{2\pi r}\right)^2 \tag{6.10}$$

with the wavelength λ_0 as the critical wavelength of a perturbation at the limit of stability. As an approximation this wavelength at a dendrite tip can be identified with the marginally stable wavelength λ_0 of a planar front, which follows from the Mullins-Sekerka theory (section 6.1). For the purely solutal case in Lipton et al. [9], the criterion is given as the product of the solutal diffusion length l_D and the capillary length d_0, which are each nondimensionalized by division through the tip radius

$$\sigma^* = \frac{2d_0 l_D}{r^2}. \tag{6.11}$$

For the case of solutal diffusion, the diffusion length is defined by the inter-diffusion coefficient in the liquid and the velocity as

$$l_D = \frac{D^{liq}}{v}. \tag{6.12}$$

Furthermore an expression for the capillary length is provided by

$$d_0 = \frac{\Gamma}{m\Delta c} \tag{6.13}$$

with the liquidus slope m, the difference of the equilibrium concentrations Δc and the Gibbs-Thomson coefficient Γ (defined in equation 6.4). One way to determine the marginal stability criterion σ^* is given by the rigorous microsolvability theory [38]. In the present thesis an alternative approach of fitting σ^* over the resulting velocities and radii from phase-field simulations is applied, which is described in chapter 14.3.2.

7 Description of the phase-field model

The phase-field method has evolved as a convenient technique to simulate the evolution of systems with complex geometries such as microstructural transformations. In the course of the current century it became a wide-spread model in the field of computational material science and is today used in many fields of research, such as solidification, biology, geosciences or engineering mechanics. The basic concept of the model is to use diffusive interfaces, which was already described in the 1950s by Ginzburg and Landau [39] and also by Cahn and Hilliard [40]. In the models of Cahn Hilliard type the diffuse interfaces represent real physical interfaces. Due to this equivalence, only simulations on small length scales can be performed with that class of models. To overcome such restrictions of possible length scales and still make use of the computational advantages of diffusive interfaces, models have been developed, which approximate the physics of sharp phase boundaries by abstract interfaces. With the increase of computer capacities at the end of the 20th century, it became possible to perform computational tasks in a reasonable amount of time, which laid the ground for the emergence of the phase-field method. The first applications for this method are from the field of solidification, for instance early results of dendritic growth were achieved by Kobayashi [41]. For the description of complex microstructures appearing in reality, multi-phase and multi-component models were introduced by Steinbach et al. [42] and by Nestler et al. [43]. These phase-field models are based on a free energy functional, whereby the free energy density is derived through an interpolation of the bulk free energy densities of the individual phases. For such an interpolation of the free energy densities, Kim [44] pointed out, that an excess energy arising from the variation of the grand potential across the interface, contributes to the interfacial energy. This excess energy increases with the difference between the equilibrium compositions of two phases and results in a reduced equilibrium width of their interface. The only way to treat systems with remarkably different equilibrium compositions with such kind of phase-field models is to simulate on smaller length scales. As a consequence, the equilibrium interface widths can be much smaller than the smallest morphological feature, which needs to be resolved. These restrictions can be overcome by the use of individual concentration fields for all phases, like it is suggested in [44, 45, 46], such that one is able to simulate on length scales interesting in solidification. An alternative to the use of separate concentration fields is given by the model introduced by Plapp et al. [47], which is based on a grand potential functional. The model used in this thesis [7] follows the same approach and is described in chapter 7.2.

7.1 General phase-field concepts

Simulations from the scope of material science must describe the physical quantities relevant for the particular process under study, such as temperature or concentration. This can be realized by a grid of computational cells, which store the scalars or tensors quantifying the considered fields. However, most materials are non-uniform on the microstructural level and consist of different phases, which are regions with homogeneous physical properties. As the material properties at a certain position are often strongly dependent on which of the phases is present, the necessity for modeling their location and evolution arises. In contrast to sharp interface models, for which the interfaces have to be tracked explicitly, the phases are described in the context of the phase-field method by continuous order parameters. Within the formulation used in this thesis, that description is implemented by the phase-field vector $\phi = \{\phi_\alpha\}_{\alpha=1}^{P}$. This vector includes the order parameters ϕ_α describing every of the phases α included in a system of P phases with a value ranging from zero to one. A value of zero hereby indicates, that the specific phase is not existing in the volume represented by the respective computational cell. In opposite a value of one determines, that at this position only the particular bulk phase is present. This condition is assured by the constraint $\sum_{\alpha=1}^{P} \phi_\alpha = 1$, which has to be fulfilled in every computational cell. The constraint is also valid for regions of coexisting phases, which are called interface and represent the phase boundaries with values of ϕ_α between zero and one. These parts of the domain are stretched smoothly about several computational cells and are not up to scale with the size of the phase boundaries in reality. An illustration of the phase-field concept can be found in figure 7.1.

This continuous description brings with it the advantage, that the interface does not have to be tracked explicitly, but is given implicitly by the order parameter.

Due to the artificial interface widths however the requirement arises, to emulate the behavior of sharp boundaries by appropriately interpolating the physical properties in the diffuse interface region. For a property Z, this is usually realized by weighting the bulk properties Z_α with a function $h_\alpha(\phi)$ and calculating the sum:

$$Z = \sum_{\alpha=1}^{P} Z_\alpha h_\alpha(\phi). \tag{7.1}$$

Possible choices for the interpolation function are plotted in figure 7.2. As exclusively simulations including no more than two phases are carried out in the scope of this thesis, only interpolation functions for two different phases are presented in the following. In doing so, it is important that the functions fulfill the summation property

$$\sum_{\alpha=1}^{P} h_\alpha(\phi) = 1. \tag{7.2}$$

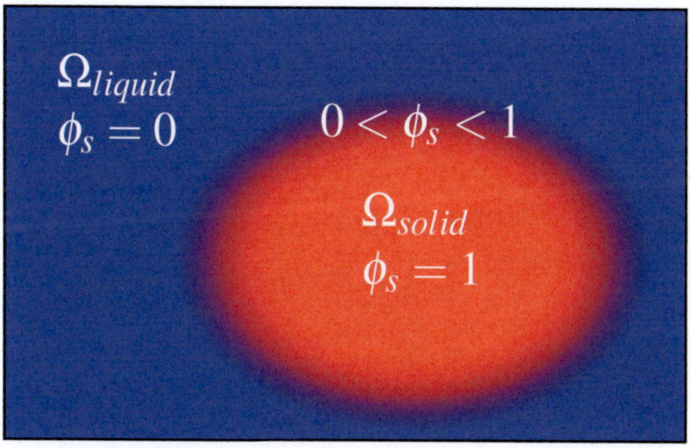

Figure 7.1: Illustration of the phase-field concept. The liquid part of the domain Ω_{liquid} (colored blue) is indicated by $\phi_s = 0$, with ϕ_s as the order parameter of the solid phase, having a value of one in the solid region Ω_{solid} (colored red). In between the solid and liquid an interfacial region exists with values of ϕ_s between zero and one.

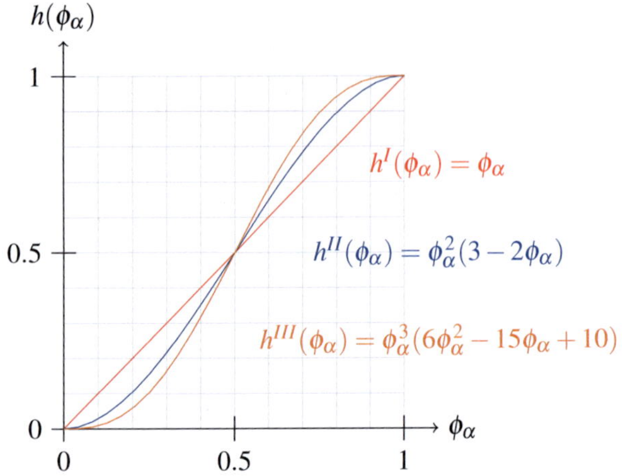

Figure 7.2: Different interpolation functions $h(\phi_\alpha)$ plotted in the domain of definition $[0,1]$.

This is the case for the simplest type of interpolation

$$h_\alpha^I(\phi) = h^I(\phi_\alpha) = \phi_\alpha, \tag{7.3}$$

but as this function is not continuously differentiable at the junctions to the bulk phases, mostly formulations of higher order are chosen, like

$$h_\alpha^{II}(\phi) = h^{II}(\phi_\alpha) = \phi_\alpha^2(3-2\phi_\alpha) \tag{7.4}$$

or for example

$$h_\alpha^{III}(\phi) = h^{III}(\phi_\alpha) = \phi_\alpha^3(6\phi_\alpha^2 - 15\phi_\alpha + 10). \tag{7.5}$$

To calculate the result for a certain phase α, these interpolation functions only take into account the phase-field parameter ϕ_α of the respective phase. For the simulations of this thesis, including solely two phases, the summation condition is fulfilled with these formulations. If more than two different phase-fields are used, one has to choose other formulations to achieve an appropriate interpolation.

7.2 Grand potential model

All simulation results presented in this thesis are derived with the quantitative phase-field model formulated by Choudhury et al. [7]. As an extensive documentation about the derivation and testing of the model can be found in the PhD thesis of Abhik Choudhury [48], the model is not described on the same level of detail. The model is based on a grand potential functional, given as the volume integral

$$\Omega(T,\boldsymbol{\mu},\boldsymbol{\phi}) = \int_V \left(\Psi(T,\boldsymbol{\mu},\boldsymbol{\phi}) + \left(\varepsilon a(\boldsymbol{\phi},\nabla\boldsymbol{\phi}) + \frac{1}{\varepsilon} w(\boldsymbol{\phi})\right)\right) dV, \tag{7.6}$$

whereby ε is a parameter related to the width of the interface. The bulk part is hereby represented by the grand potential density Ψ, which is dependent on the phase-field parameters, the temperature T and the vector $\boldsymbol{\mu} = \{\mu_i\}_{i=1}^{K-1}$ encapsulating the $K-1$ independent chemical potentials for a system comprising of K components. The interfacial contributions consist of the gradient energy density a and the surface energy potential w. In doing so, the part of the potential energy lets the bulk phases be energetically favorable, which causes the interface to get narrower. In contrast the gradient energy density widens the interface, as it reduces gradients of the phase-field parameters. The interplay of these two opposing terms leads to the formation of stable interfaces having finite widths, whereas a driving force for phase transformations is caused by differences in the grand potentials. Before the model gets explained in greater detail, a short excursus about the used definitions is given in the following.

7.2.1 Definitions used in the grand potential model

In literature different notations and definitions for physical quantities can be found depending on the author and which branch of science he belongs to. To avoid any possibility of confusion, the definitions used in the grand potential formulation are distinguished from other definitions for the same terms in the following.

Generally, concentrations are defined as a chemical composition per volume, like the molar concentration N_i/V. However, in literature the term is sometimes also used for the mole fraction or mass fraction, for example in the book of Kurz and Fisher [34] or in many phase-field publications [49, 43, 50]. This definition is justified under the assumption of constant and equal molar volumes V_m for all phases, because that results in a linear relationship $\frac{N_i}{N} = \frac{N_i}{V} \times V_m$ between the volume related quantities and the mole fractions, as pointed out by Heulens et al. [51]. In consistency with the notation of the grand potential model [7] the concentration is defined in the following as the fraction $c_i = \frac{N_i}{N}$ and can be equated with the mole fractions x_i used in the previous chapters. In consequence of this definition, the concentration is a conserved quantity and has to fulfill the constraint

$$\sum_{i=1}^{K} c_i = 1. \tag{7.7}$$

This constraint can be used to explicitly define the composition of a system of K components with only $K-1$ independent concentrations, which are given by the vector $\mathbf{c} = \{c_i\}_{i=1}^{K-1}$. To be able to distinguish between the dependent and independent concentrations, the order of the components is defined in this thesis, such that the dependent component is always the K-th element and thus c_K is the dependent concentration. In doing so, the order of the components does not have to be the same as the alphabetic order and any concentration of a system can be defined as the dependent one.

The thermodynamic quantities used in the grand potential model are derived from the free energy density f. In [7] the free energy density is defined as $f = G_m/V_m$ with G_m as the free energy of a system consisting of one mole of particles and V_m as the molar volume. As set out in section 3.1, there is a difference between Helmholtz and Gibbs free energy, which is a constant for the conditions of constant pressure and volume assumed in the phase-field model. Because only derivatives of the free energies and differences between them contribute to the evolution of the fields, it is thus justified to identify the free energy in the context of the grand potential model either with the Gibbs energy or with the Helmholtz energy.

In general the chemical potential μ_i is defined as the partial derivative of the Gibbs energy with respect to the number of particles (eq. 3.8). In this model a different definition for the chemical potential as partial derivative of the free energy density with respect to the independent concentrations $\mu_i = \frac{\partial f(\mathbf{c})}{\partial c_i}$ is chosen. In doing so, the free energy densities have to be expressed as functions of solely the independent concentrations. If the formulations also include the dependent concentration c_K, it has to be replaced by $1 - \sum_{i=1}^{K-1} c_i$ in order to calculate the partial derivatives giving the chemical potentials. This definition shall be explained by a simple example, for which the free energy of a phase in a binary system A-B at constant temperature is defined by the concentrations of the two components as

$$f^\alpha(c_A, c_B) = f_A^\alpha c_A + f_B^\alpha c_B + RT\left(c_A \ln(c_A) + c_B \ln(c_B)\right) + f_{const}^\alpha. \tag{7.8}$$

If the concentrations are treated as independent variables, the derivative of the free energy density with respect to c_A is given by $f_A^\alpha + RT\left(1 + \ln(c_A)\right)$ and the derivative with respect to c_B by $f_B^\alpha + RT\left(1 + \ln(c_B)\right)$. This derivative corresponds to the chemical potential definition used in the phase-field model of Nestler et al. [43]. To derive the chemical potentials for the grand potential model, a single independent concentration can be defined as $c = c_B$ and with $c_A = 1 - c$ the derivative with respect to c and thus $\mu^\alpha(c)$ is given as

$$\begin{aligned} \frac{\partial\left(f_A^\alpha(1-c) + f_B^\alpha c + RT\left((1-c)\ln(1-c) + c\ln(c)\right) + f_{const}^\alpha\right)}{\partial c} = \\ f_B^\alpha - f_A^\alpha + RT \ln\left(\frac{c}{1-c}\right). \end{aligned} \tag{7.9}$$

Instead of replacing the dependent concentration c_K, the chemical potentials can also be derived as

$$\mu_i^\alpha(T, \mathbf{c}, c_K) = \frac{\partial f^\alpha(T, \mathbf{c}, c_K)}{\partial c_i} - \frac{\partial f^\alpha(T, \mathbf{c}, c_K)}{\partial c_K}. \tag{7.10}$$

Applied to the example system this leads to

$$\frac{\partial f^{\alpha}(c_A,c_B)}{\partial c_B} - \frac{\partial f^{\alpha}(c_A,c_B)}{\partial c_A} = f_B^{\alpha} - f_A^{\alpha} + RT\ln\left(\frac{c_B}{c_A}\right), \tag{7.11}$$

which is identical to the previous result for the chemical potential.

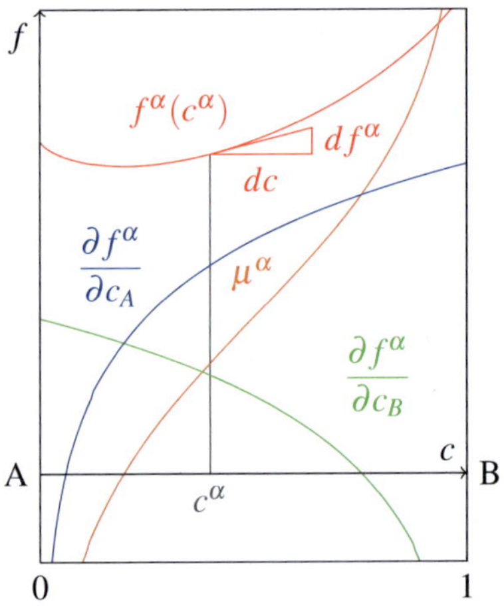

Figure 7.3: The definition of the chemical potential in the context of the grand potential model is illustrated for the phase α of a binary system. For this case the chemical potential μ^{α} is given as the derivative of the free energy density f^{α} with respect to the independent concentration $c = c_B$. The diagram also shows the curves of the two derivatives, which are derived treating the concentrations as independent variables.

Having a look on the free energy density diagram, the chemical potentials can be identified as the slopes of the free energy density curves at a certain composition. In figure 7.3 the chemical potential of phase α at the concentration c^{α} is given by the tangent to f^{α} at this composition and acts as a factor of proportionality between df^{α} and dc. The definition of the chemical potentials in the grand potential model can also be schematically visualized similar to the illustration given in figure 3.1. For the two binary configurations drawn in figure 7.4 the chemical potential is related to their difference in the free energy $dG = Vdf$. In contrast to the previous definition (eq. 3.8), the number of atoms of the system is kept constant and the change in energy is caused by the replacement of one atom by an atom of the other species.

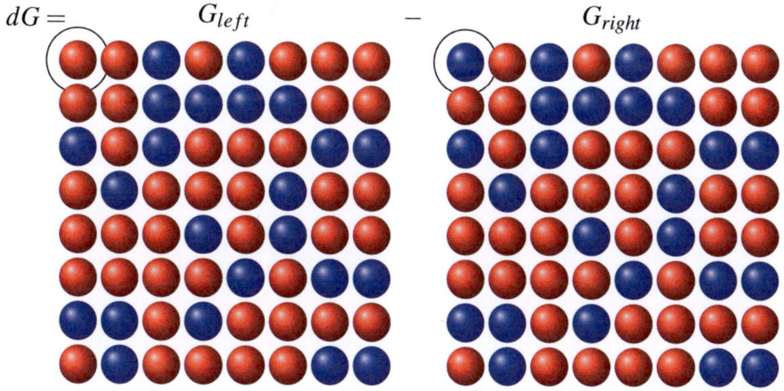

Figure 7.4: Visualization of the definition of the chemical potential used in the grand potential model. The right side shows the original system consisting of atoms A (red) and B (blue). On the left side one additional atom of A is added to the system at the marked position in the upper left corner. In contrast to the addition shown in figure 3.1 the number of atoms of the system is conserved. The atom of type B, that was originally located at the marked position, was removed from the system.

In general, the grand potential is given as $G - \sum_{i=1}^{K-1} \mu_i N_i$, with G being the free energy of a system consisting of N particles, N_i as the number of particles of component i and $\mu_i = \frac{\partial G}{\partial N_i}$. By applying the relation $N_i = c_i N$ with the concentration c_i of component i given as a mole fraction and using the definition $\mu_i = \frac{\partial f}{\partial c_i}$, the grand potential for one mole of particles can be expressed as

$$\Psi_m = G_m - V_m \sum_{i=1}^{K-1} \mu_i c_i \tag{7.12}$$

under the assumption that the molar volume is the same for all particles. From this definition, the grand potential density $\Psi = \Psi_m / V_m$ used in this phase-field model can be interpolated between the grand potential densities Ψ^{α} for each phase α and follows as

$$\Psi(T,\boldsymbol{\mu},\phi) = \sum_{\alpha=1}^{P} \Psi^{\alpha}(T,\boldsymbol{\mu})\, h_{\alpha}(\phi) \tag{7.13}$$

by using the interpolation function $h_{\alpha}(\phi)$ and with

$$\Psi^{\alpha}(T,\boldsymbol{\mu}) = f^{\alpha}(T,\mathbf{c}^{\alpha}(T,\boldsymbol{\mu})) - \sum_{i=1}^{K-1} \mu_i c_i^{\alpha}(T,\boldsymbol{\mu}). \tag{7.14}$$

The vector $\mathbf{c}^{\alpha} = \{c_i^{\alpha}(T,\boldsymbol{\mu})\}_{i=1}^{K-1}$ encapsulates the phase concentration functions $c_i^{\alpha}(T,\boldsymbol{\mu})$, which are defined as the inverse of the functions $\mu_i^{\alpha}(T,\mathbf{c}) = \dfrac{\partial f^{\alpha}(T,\mathbf{c})}{\partial c_i}$ and for which the following equality holds:

$$c_i(T,\boldsymbol{\mu},\phi) = \sum_{\alpha=1}^{P} h_{\alpha}(\phi) c_i^{\alpha}(T,\boldsymbol{\mu}). \tag{7.15}$$

A graphical interpretation of the grand potential densities can be identified in the diagram of the free energy densities from the binary example system introduced before. In figure 7.5 the grand potential density for c^{α} is given by the intersection point of the f-axis with the tangent to f^{α} at c^{α}. As phase transformations in alloys proceed in order to minimize the grand potential densities, their difference $\Delta\Psi$ acts as the driving force for such transformations.

Evolution equation of the phase-field

Due to the condition, that the grand potentials tend towards a minimum to achieve a state of equilibrium, the functional defined in equation 7.6 can be used to derive the evolution equation for the phase-fields. As an outcome of taking the variational derivatives of the functional, the following equation can be used to calculate the temporal change of the phases in each computational cell:

$$\begin{aligned}\tau\varepsilon\frac{\partial\phi_\alpha}{\partial t} =&\varepsilon\left(\nabla\cdot\frac{\partial a(\phi,\nabla\phi)}{\partial\nabla\phi_\alpha}-\frac{\partial a(\phi,\nabla\phi)}{\partial\phi_\alpha}\right)-\\ &\frac{1}{\varepsilon}\frac{\partial w(\phi)}{\partial\phi_\alpha}-\frac{\partial\Psi(T,\mu,\phi)}{\partial\phi_\alpha}-\lambda.\end{aligned} \tag{7.16}$$

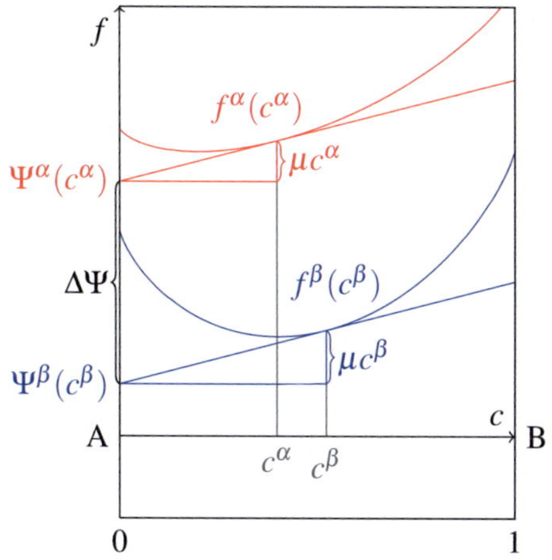

Figure 7.5: Definition of the grand potential densities in the context of the used phase-field model. For each phase the grand potential density at a certain composition is given by the intersection of the f-axis with the tangent to the free energy density curve at this composition. The driving force for the transition from phase α to phase β is given by the difference $\Delta\Psi$ between the grand potential densities of the two phases.

The Lagrange parameter λ takes care, that by doing so the constraint $\sum_{\alpha=1}^{P}\phi_\alpha = 1$ is fulfilled. The relaxation constant τ governs the kinetics of the phase transformations and needs to be chosen in a certain manner, as explained further below. A formulation for the gradient energy density is given by

$$a(\phi, \nabla\phi) = \sum_{\substack{\alpha,\beta=1 \\ (\alpha<\beta)}}^{P,P} \sigma_{\alpha\beta} \left[a_c\left(q_{\alpha\beta}\right)\right]^2 |q_{\alpha\beta}|^2 \ , \tag{7.17}$$

where $\sigma_{\alpha\beta}$ is the surface energy. The normal vector to the interface between the phases α and β is given as $q_{\alpha\beta} = \left(\phi_\alpha \nabla \phi_\beta - \phi_\beta \nabla \phi_\alpha\right)$ and the type of anisotropy of the phase boundaries is described by $a_c\left(q_{\alpha\beta}\right)$. To model cubic anisotropy, the formulation

$$a_c\left(q_{\alpha\beta}\right) = 1 - \delta_{\alpha\beta}\left(3 \mp 4\frac{|q_{\alpha\beta}|_4^4}{|q_{\alpha\beta}|^4}\right) \tag{7.18}$$

can be used in equation 7.17 with the expressions $|q_{\alpha\beta}|_4^4 = \sum_i^d (q_{\alpha\beta})_i^4$ and $|q_{\alpha\beta}|^4 = \left[\sum_{i=1}^d (q_{\alpha\beta})_i^2\right]^2$, which include d as the number of dimensions and $\delta_{\alpha\beta}$ as the amplitude of the surface energy anisotropy. Throughout this thesis a double obstacle type is applied for the surface energy potential, which is only defined on the Gibbs-Simplex. It has the form

$$w(\phi) = \begin{cases} \dfrac{16}{\pi^2} \displaystyle\sum_{\substack{\alpha,\beta=1 \\ (\alpha<\beta)}}^{P,P} \sigma_{\alpha\beta}\phi_\alpha\phi_\beta & \text{if} \quad \phi_\alpha, \phi_\beta \geq 0 \text{ and } \phi_\alpha + \phi_\beta = 1 \\ \infty & \text{else} \end{cases} \tag{7.19}$$

and is plotted in figure 7.6.

For the derivation of the kinetic factor τ one has to consider the relevant time scales of the processes to be simulated. Compared to phase transformations happening at the microscale, the relaxation of the interface, which is a phenomenon at the atomic level, happens in a negligibly small amount of time. For this reason the kinetic factor should be chosen such that the phase-field reacts to the applied driving forces without a delay. An equation for the relaxation constant is derived by a thin interface analysis for the used obstacle potential in [7]. For the derivation, the phase-field and chemical potentials are written as powers of the parameter ε for an inner region, which is characterized by rapid changes of the fields, and the outer regions, where the changes are happening more slowly. By comparison of the inner and outer solutions an equation for vanishing kinetics of an interface between two phases α and β is determined as

$$\tau = \varepsilon \left[c_i^\beta \left(T, \boldsymbol{\mu}_{eq}\right) - c_i^\alpha \left(T, \boldsymbol{\mu}_{eq}\right) \right]_{1\times(K-1)} \left[D_{ij}^\beta \frac{\partial c_i^\beta \left(T, \boldsymbol{\mu}^0\right)}{\partial \mu_j} \right]_{(K-1)\times(K-1)}^{-1}$$
$$\left\{ c_j^\beta \left(T, \boldsymbol{\mu}^0\right) - c_j^\alpha \left(T, \boldsymbol{\mu}^0\right) \right\}_{(K-1)\times 1} \times (\tilde{M} + \tilde{F}), \tag{7.20}$$

with $\{\}$ representing a vector and $[]$ a matrix. The formula includes the solvability integrals $\tilde{M}$ and $\tilde{F}$, which are given for the applied obstacle potential in table 7.1 in dependence of the interpolation functions $h(\phi_\alpha)$.

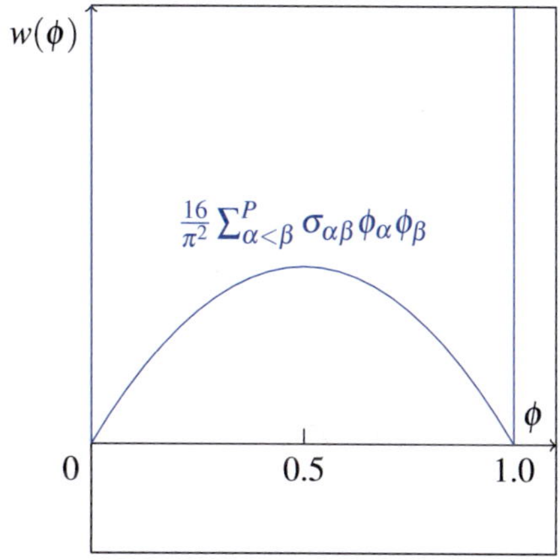

Figure 7.6: Plot of the double obstacle potential in the domain of definition $[0, 1]$.

For binary alloys $\boldsymbol{\mu}^0$ stands for the macroscopic interfacial chemical potential in the sharp interface limit. To determine the exact value of $\boldsymbol{\mu}^0$ for every computational cell and every time step requires to calculate the average value of the chemical potential over the interface. As this is computationally very expensive, there are two usual approaches of approximating $\boldsymbol{\mu}^0$. The first one is to identify it with the equilibrium chemical potentials μ_{eq} before the start of the simulation and to use static kinetic coefficients. This is an adequate approximation for small undercoolings, for which the interfacial chemical potentials are close to the equilibrium values. The second method is to identify μ^0 with the local value of μ, which is currently present in the simulation domain at the respective position. This dynamic approach is more accurate, as it does not assume values close to equilibrium for all parts of the domain. As this method requires the calculation of τ for every cell and time step, it involves a higher computational effort. Nevertheless it is used for all simulations of solidification, presented in this thesis, which only include one solid and one liquid phase. For the case of anisotropic surface energies a similar approach as in [52] can be used to obtain vanishing interface kinetics in all normal directions. The relaxation constant for the anisotropic case is then chosen as the product

$$\tau\left(q_{\alpha\beta}\right) = \tau^{\alpha\beta}\left(a_c\left(q_{\alpha\beta}\right)\right)^2. \tag{7.21}$$

The applied $\tau^{\alpha\beta}$ can be calculated with equation 7.20 and $a_c\left(q_{\alpha\beta}\right)$ with equation 7.18.

Table 7.1: Values of the solvability integrals for the employed interpolation polynomials

Potential	$\tilde{M}$	$\tilde{F}$
$h^{II}(\phi_\alpha) = \phi_\alpha^2(3-2\phi_\alpha)$	0.063828	0.158741
$h^{III}(\phi_\alpha) = \phi_\alpha^3\left(10-15\phi_\alpha+6\phi_\alpha^2\right)$	0.052935	0.129288

Evolution equation for the chemical potentials

In addition to the evolution equations for the phase-fields, the grand potential formulation also includes the solving for the chemical potentials. In [7] the evolution equation for the $K-1$ independent chemical potentials is given by:

$$\left\{\frac{\partial \mu_i}{\partial t}\right\} = \left[\sum_{\alpha=1}^{P} h_\alpha(\phi)\frac{\partial c_i^\alpha(T,\boldsymbol{\mu})}{\partial \mu_j}\right]_{ij}^{-1}$$
$$\left\{\nabla\cdot\left(\left(\sum_{j=1}^{K-1} M_{ij}(\phi)\nabla\mu_j\right) - \vec{j}_i\right) - \sum_{\alpha=1}^{P} c_i^\alpha(T,\boldsymbol{\mu})\frac{\partial h_\alpha(\phi)}{\partial t}\right\}. \tag{7.22}$$

In the region of the interface, the mobilities $M_{ij}(\phi)$ are given by an interpolation between the phases by

$$M_{ij}(\phi) = \sum_{\alpha=1}^{P} M_{ij}^\alpha g_\alpha(\phi) \tag{7.23}$$

with the interpolation function $g_\alpha(\phi)$. In general this function can be chosen individually, but throughout this thesis the same formulation as for $h_\alpha(\phi)$ is applied. The mobilities of the phases M_{ij}^α are given by the inter-diffusivities D_{ij}^α and a thermodynamic factor, which can be calculated as the derivative of the phase concentrations with respect to the chemical potential:

$$M_{ij}^\alpha = D_{ij}^\alpha \frac{\partial c_i^\alpha(T,\boldsymbol{\mu})}{\partial \mu_j}. \tag{7.24}$$

The previous equation includes an anti-trapping current $\vec{j}_i$, which is applied to simulate processes like solidification with markedly different diffusivities in the solid and liquid. In such cases the phase-field model leads to an artificial solute trapping effect, which results in incorrect concentrations of the forming solid phase. To match with the sharp interface solution, the method of an additional flux, called the anti-trapping current, was proposed by Karma [53]. The appropriate mechanism and derivation of the anti-trapping term for the case of the obstacle potential is given in [7]. The anti-trapping current $\vec{j}_i$ is an additional interfacial flux, oriented from the solid phase α to the liquid phase l:

$$\vec{j}_i = -\frac{\pi \varepsilon g_\alpha\left(\phi_\alpha^0\right)\left(1-h_\alpha\left(\phi_\alpha^0\right)\right)}{4\sqrt{\phi_\alpha^0\left(1-\phi_\alpha^0\right)}}\left(c_i^l\left(T,\mu^0\right)-c_i^\alpha\left(T,\mu^0\right)\right)\frac{\partial \phi_\alpha}{\partial t}\frac{\nabla \phi_\alpha}{|\nabla \phi_\alpha|}. \quad (7.25)$$

In this equation the ϕ_α^0 denotes the solution of the phase-field equation in the asymptotic analysis at zero order in ε. In the same manner as for the dynamical calculation of τ the values for μ^0 and ϕ_α^0 can be approximated with the current local values from the simulation domain.

Modeling of varying temperatures

If directional solidification is realized for example with a bridgman furnace, the temperature distribution inside a crucible is controlled by its motion relative to a heater. For modeling directional solidification, the so called “frozen temperature approximation” is often applied, as for example in [54]. This model underlies the assumption, that the temperature inside the crucible is totally defined by the process conditions, given by the velocity and steepness of the imposed temperature gradient. The temperature distribution can thus be described as a function of space and time. For a linear temperature gradient moving in positive z-direction with the velocity v, the temperature at a location within the temperature gradient can be calculated in dependence of the coordinate z and time t:

$$T(z,t) = T_0 + G(z - z_0 - vt). \quad (7.26)$$

In this equation G denotes the slope of the gradient in growth direction, with z_0 as the offset in z-direction and with T_0 as the base temperature. An illustration of the gradient is given by figure 7.7. Such a gradient is for example applied in a simulation study about ternary eutectic growth by Hötzer et al. [55].

7.2.2 Temperature dependence of the chemical potential

The previous evolution equation 7.22 for the chemical potentials is valid for the case of constant temperature. Under the condition of varying temperatures, the need for a modified formulation arises, which can be elucidated by the following exemplary setting. For a simple binary system consisting of only one single phase the diffusivity is chosen to be zero. The boundary conditions are such, that no mass transfer can happen between the system and its surrounding. Initially the composition is uniformly distributed inside the phase and hence no concentration gradients exist. Similarly, the temperature is the same at any position of the system. Without nucleation, no phase transformations can occur and with zero diffusivity and uniform concentrations, there is no cause for a change of concentration. For the grand potential model, the initial concentration field has to be transformed into a corresponding chemical potential field. Now the temperature is varied by the same rate everywhere, such that no temperature gradients occur. According to the evolution equation 7.22, the equally distributed chemical potential field remains constant for such a setup. A function $c_i(T, \boldsymbol{\mu})$ can be derived, for instance from the CALPHAD formulations, to recalculate the concentrations from the chemical potentials. Because such a function is generally dependent on temperature, the resulting concentrations at a later time differ from the initial ones since the temperature changed. This violates the rule of mass conservation, as the overall concentration of the system is not allowed to change with time.

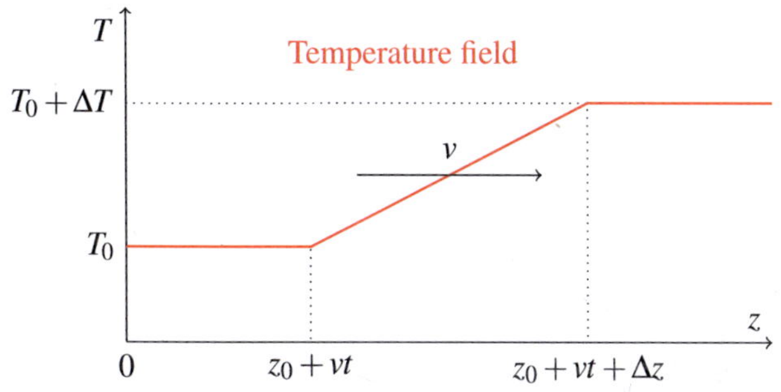

Figure 7.7: A moving temperature field for the modeling of directional solidification. The slope of the gradient is given as $G = \Delta T / \Delta z$.

A solution to this miscalculation can be derived starting from the temporal evolution of the concentrations and making use of the equality from equation 7.15:

$$\left\{\frac{\partial c_i(T,\mu,\phi)}{\partial t}\right\} = \left[\sum_{\alpha=1}^{P} h_\alpha(\phi)\frac{\partial c_i^\alpha(T,\mu)}{\partial \mu_j}\right]_{ij} \left\{\frac{\partial \mu_j}{\partial t}\right\} + \left\{\sum_{\alpha=1}^{P} c_i^\alpha(T,\mu)\frac{\partial h_\alpha(\phi)}{\partial t} + \sum_{\alpha=1}^{P} h_\alpha(\phi)\left(\frac{\partial c_i^\alpha(T,\mu)}{\partial T}\right)\frac{\partial T}{\partial t}\right\} = \left\{\nabla\cdot\left(\left(\sum_{j=1}^{K-1} M_{ij}(\phi)\nabla\mu_j\right) - \vec{j}_i\right)\right\}. \tag{7.27}$$

Keeping in mind, that the number of independent concentrations and chemical potentials is the same, the previous equation can be rearranged to the extended formula for the temporal evolution of the chemical potential

$$\left\{\frac{\partial \mu_i}{\partial t}\right\} = \left[\sum_{\alpha=1}^{P} h_\alpha(\phi)\frac{\partial c_i^\alpha(T,\mu)}{\partial \mu_j}\right]_{ij}^{-1} \left\{\nabla\cdot\left(\left(\sum_{j=1}^{K-1} M_{ij}(\phi)\nabla\mu_j\right) - \vec{j}_i\right) - \sum_{\alpha=1}^{P} c_i^\alpha(T,\mu)\frac{\partial h_\alpha(\phi)}{\partial t} - \sum_{\alpha=1}^{P} h_\alpha(\phi)\left(\frac{\partial c_i^\alpha(T,\mu)}{\partial T}\right)\frac{\partial T}{\partial t}\right\}. \tag{7.28}$$

The last term in the equation accounts for the change in temperature. It is formulated as a weighted sum over all partial derivatives of the phase concentrations with respect to temperature multiplied with the temporal change of temperature. For a temperature gradient as defined in equation 7.26, the temporal change of temperature is constant and given by $\partial T/\partial t = -Gv$. A description of the extended equation and the derivation of the required thermodynamic properties for the case of a temperature variation can be found in Choudhury et al. [56].

7.2.3 Boundary conditions

For the fields of the phase parameters and the chemical potentials, different boundary conditions are used throughout this thesis. The first one models periodic conditions for a pair of opposite boundaries. It is implemented by copying the values of a specific field from the outermost layers inside the actual simulation domain into the corresponding boundary layers on the opposite side. In the same way, the outer layers from the opposite side get transferred. The second type of conditions used in this thesis are Neumann boundary conditions, for which the directional derivatives normal to the domain boundaries are set to zero. This is implemented by copying the outermost layers of the actual simulation domain into the adjacent boundary layers. The domain boundaries act as axis of symmetry for this condition and it is therefore also called mirror condition.

A third type of boundary condition is used in simulations of directional solidification, which sets the chemical potentials in an extrapolative way to reproduce the diffusion profile outside of the domain. This boundary condition includes the assumption, that the chemical potential fields of all components i in growth direction z are given by a function

$$\mu_i(z) = \mu_i^\infty + A_i \exp(a_i z), \tag{7.29}$$

where μ_i^∞ denotes the chemical potential far away from the front. The values of the chemical potential field close to the boundary are used to extrapolate the value inside the boundary cell in normal direction via the parameters A_i and a_i.

Part III

Coupling of the grand potential model with CALPHAD databases

8 Overview of coupling methods

To achieve realistic simulation results, accurate data about the thermodynamic properties of the material system under investigation is a prerequisite. These days CALPHAD databases (see section 5.1) are often utilized in phase-field studies as a source for these essential material properties and have a significant influence on the course of the simulated phase transitions. For reducing the computational demands of large scale simulations, it is most of the time convenient to approximate the required functions in the region of interest. An early example for this practice of coupling phase-field solvers directly with thermodynamic databases is the approach from Grafe et al. [57] from the beginning of the new millennium. In their framework the information about the Gibbs energies and chemical potentials is applied to determine the driving forces at the diffuse interface and for obtaining the diffusion matrix. The practical implementation of the CALPHAD coupling is realized via the application programming interface of commercial thermodynamic software. Another example for the successful usage of thermodynamic databases is the simulation of precipitation of Ni-base super-alloys by Zhu et al. [58]. Furthermore Siquieri et al. [59] use simplified expressions for the Gibbs energies from CALPHAD in their work on peritectic growth. They apply thermodynamic and kinetic data from a composition and temperature dependent free energy description including multiple sub-lattices. Qin and Wallach [60] also use values for the molar Gibbs energy and chemical potential from thermodynamic databases to calculate the phase-field driving forces and apply it for simulating the solidification of aluminum-silicon alloys. They optimize their linking to a commercial thermodynamic software, to reduce the communication effort. In the coupling method of Kobayashi et al. [61], computational effort is reduced by precalculating the tie-line concentrations from CALPHAD data in an initial step. The stored tie-line data can then be called during phase-field simulations, like for dendritic growth of an Al-Si-Mg alloy. The phase-field formulation of Eiken et al. [46] requires to solve numerically for quasi-equilibrium of the free energies. To reduce the computational effort they suggest a similar optimization as in the previously cited article. In their approach the phase-field routines can be decoupled from the calculation of the quasi-equilibrium data by storing it in a periodically updated buffer. A strategy to incorporate the sublattice formulation from the CALPHAD method into the phase-field context can be found in [62].

As described in section 7.2, the grand potential model from Choudhury and Nestler [7] requires functions for the grand potentials, the phase concentrations and their derivatives. The CALPHAD method does not provide explicit functions for these thermodynamic quantities. Therefore appropriate formulations need to be found for them, such that the unknown coefficients can be derived from the available Gibbs energy data. It is hereby beneficial from a computational point of view to use purpose-built expressions, which can efficiently be evaluated by the solver internally, thus preventing a loss of performance occurring from communication to external software. So far, the use of CALPHAD data for the grand potential model has been realized by the approach of precalculating the free energies as polynomial approximations of adequate order. As deviations from equilibrium are generally very small for phase-field applications, hence fitting the required thermodynamic quantities around the equilibrium compositions is sufficient. A successful application of this procedure for the simulation of the eutectoid transformation in steel can be found in [63]. The same procedure is also employed in the study on solidification of the system aluminum-magnesium-silicon presented in this thesis (in section 14). A detailed discussion about the topic of CALPHAD coupling in the context of the grand potential model can be found in the following chapters.

9 Overview of system setting data

The CALPHAD databases provide Gibbs energy descriptions, which are suited to accurately reproduce the respective phase-diagrams in their entirety. Since the calculation of phase-diagrams is not a challenge for today's computer systems, the focus of this task lies on the accuracy of the description and not on the computational effort. The utilization of datasets for phase-field simulations is however a different case of application, which entails a modification of the objectives. Thus, the complexity of the required calculations has to be taken into account for being able to perform simulations in an efficient manner. At the same time, the thermodynamic modeling should not lead to a loss of the characteristics, which are important for the phenomena under study. It has to be ensured, that the relevant details of the phase-diagram are preserved and also the right values of parameters dependent on thermodynamic quantities, like the Gibbs-Thomson coefficient, get reproduced. The following listing gives an overview of the aspects, which have to be considered in the context of CALPHAD coupling via approximated thermodynamic functions:

- The functions can be fitted by the least squares method over a composition range or they can be constructed to perfectly match at a particular composition, as it is the case for Taylor approximations. The latter can be used to exactly reproduce an equilibrium between phases and is applied to the description of a stoichiometric phase in section 11.4.
- For the choice of the coupling approach, the number of components plays an important role. Throughout this thesis only binary and ternary systems (as in section 14.2.2) are treated. Nevertheless the methodology can be generalized to multi-component systems with four or more components. An additional aspect is the description of a ternary system as a pseudobinary system, which gets briefly addressed in section 11.5.

- The thermodynamic functions of one phase can be derived independently from each other, like it is done in the study of Al-Si (section 14.1). Alternatively the function of a single thermodynamic quantity can be approximated and used to derive the expressions for the related quantities. For example all thermodynamic functions for the phases in the study of Fe-Cu in 15 are derived from the approximations of the Gibbs energy. This is however only possible if there exists an explicit analytical transformation.
- The Gibbs energies can be approximated with different kinds of expressions. These formulations can include solely polynomials or also further terms like in the case of an ideal solution.
- A single phase from a CALPHAD dataset can be represented by two or more phases with simple Gibbs energy expressions in a phase-field setup. This can be useful if the curves exhibit two minima, like in the example for Fe-Cu in section 15.
- Thermodynamic functions can be constructed for a single temperature or they can be described as temperature dependent quantities (like in section 11.4).

These different aspects together with two variants for each case are listed in table 9.1. In the following sections, the parameter determination for different thermodynamic models and the associated trade-off between accuracy and complexity are discussed. At the example of case studies, which are also marked in the table, these aspects are treated in more detail. Regarding the last item of the listed aspects, it has to be mentioned, that the majority of simulations presented in this thesis is done for a constant temperature. For this task solely thermodynamic functions dependent on concentrations or chemical potentials are necessary, as the parameters from the CALPHAD formulations reduce to constants. The following sections about the different approximation methods therefore refer to the case of a fixed temperature. An approach to model a variation of temperatures can be found in section 11.3.

Table 9.1: Different aspects regarding the framework of CALPHAD coupling presented in this thesis. The combinations corresponding to the following case studies are marked by colored circles and represent: ❶ Al-Si fitting (sec, 14.1), ❷ Approximation of Al-Si-Mg (sec. 14.2.2), ❸ Modeling of Fe-C (sec. 11.4), ❹ Pseudobinary coupling (sec. 11.5), ❺ Fe-Cu fitting (sec. 15)

Aspects	Variants	
Number of components	Binary ❶❹❺❸	Ternary ❷
Dependency of functions	Independent functions ❶	Derived functions ❹❺❸❷
Formulation	Polynomials ❶❺❸❷	Pol. with additional terms ❹
Parameter determination	Least squares fitting ❶❹❺	Expansion around point ❸❷
Representation of phases	Single phases ❶❹❸❷	Several phases ❺
Temperature	Single temperature ❶❹❺❷	Temperature range ❸

10 Simplified formulations for the thermodynamic functions

As mentioned previously, an important issue of utilizing CALPHAD data for phase-field simulations is to minimize the computational effort required for the calculation of the thermodynamic quantities. This can be achieved by using simplified formulations for the thermodynamic functions, which are needed in order to solve the evolution equations of the specific model. Given an appropriate formulation of the free energy densities, the functions for the thermodynamic quantities of the grand potential model can be derived analytically and the unknown parameters can be calculated from the CALPHAD Gibbs energies. Possible choices for such formulations, which have already proven their applicability, are described in this section. The simplest type of formulation are polynomials with the independent concentrations as variables. Another possibility is to use polynomials with additional terms, like ideal, regular or sub-regular solutions (see also section 3.2) including logarithmic expressions from configurational entropy (given by equation 3.17 for the binary case). To solve the evolution equation of the chemical potentials 7.22, expressions for the phase concentrations are required, which can be derived from the formulations of the free energy densities, if the corresponding chemical potential functions are invertible. This is for example possible for ideal solution models or if the free energy densities are described as quadratic functions, which is addressed in the following. If the functions of the phase concentrations cannot be derived analytically from the free energy densities, they can also be approximated separately, as explained in section 11.2.

10.1 Formulation as quadratic polynomials

One way to describe the free energy densities in a simplified manner is the application of polynomials [64, 56], which are usually chosen to be quadratic. The use of linear polynomials for the free energy densities in dependence of concentration would result in constant chemical potentials, for which the inverse functions are undefined. Such a linear formulation can also represent only constant grand potential densities and therefore the simplest applicable polynomials are of degree two. For the binary case with only one independent component and constant temperature, the quadratic free energy density functions can be written as

$$ {}^{P}f^{\alpha}(c) = A^{\alpha}c^{2} + B^{\alpha}c + X^{\alpha}, \tag{10.1} $$

with the P in the pre-superscript indicating a polynomial formulation. From the above expression depending on the three coefficients A^{α}, B^{α} and X^{α}, all thermodynamic functions required for the grand potential model can be derived. For this purpose one can again calculate a function for the chemical potentials, which is given by the first derivative of the polynomial

$$ {}^{P}\mu^{\alpha}(c) = 2A^{\alpha}c + B^{\alpha}. \tag{10.2} $$

By inverting this expression, the phase concentrations can be derived as functions of the chemical potential

$$ {}^{P}c^{\alpha}(\mu) = \frac{\mu - B^{\alpha}}{2A^{\alpha}} \tag{10.3} $$

and their derivatives with respect to μ follow as

$$ \frac{\partial\, {}^{P}c^{\alpha}(\mu)}{\partial \mu} = \frac{1}{2A^{\alpha}}. \tag{10.4} $$

Finally, by inserting the expression for ${}^{P}c^{\alpha}(\mu)$ in ${}^{P}f^{\alpha}(c) - \mu c$, one can derive an expression for the approximated grand potential densities

$$ {}^{P}\Psi^{\alpha}(\mu) = X^{\alpha} - \frac{(\mu - B^{\alpha})^2}{4A^{\alpha}} = -\frac{1}{4A^{\alpha}}\mu^2 + \frac{B^{\alpha}}{2A^{\alpha}}\mu + X^{\alpha} - \frac{(B^{\alpha})^2}{4A^{\alpha}} \tag{10.5} $$

and with this all the required thermodynamic functions for the grand potential model in the case of a binary system are available. Obviously, the grand potential densities are given as quadratic polynomials depending on the chemical potentials like in equation 11.21. Instead of determining the coefficients of the free energy densities, the coefficients of the grand potential densities can thus be directly calculated from the CALPHAD data, as it is done for the system iron carbon in section 11.4.

For systems with more than two components, the free energy density can be written as

$$ {}^{P}f^{\alpha}(\mathbf{c}) = \sum_{\substack{i=1,j=1 \\ i<=j}}^{K-1} A^{\alpha}_{i,j} c_i c_j + \sum_{i=1}^{K-1} B^{\alpha}_i c_i + X^{\alpha}, \tag{10.6} $$

like in [48]. It is appropriate to write this function in matrix notation, as used in [55]:

$$ {}^{P}f^{\alpha}(\mathbf{c}) = \langle \mathbf{c}, \Xi^{\alpha}\mathbf{c}\rangle + \langle \mathbf{c}, \xi^{\alpha}\rangle + X^{\alpha} \tag{10.7} $$

with the coefficients being included in the matrix

$$ \Xi^{\alpha} = \begin{pmatrix} A^{\alpha}_{1,1} & \frac{A^{\alpha}_{1,2}}{2} & \cdots & \frac{A^{\alpha}_{1,K-1}}{2} \\ \frac{A^{\alpha}_{1,2}}{2} & A^{\alpha}_{2,2} & \cdots & \frac{A^{\alpha}_{2,K-1}}{2} \\ \vdots & \vdots & \ddots & \vdots \\ \frac{A^{\alpha}_{1,K-1}}{2} & \frac{A^{\alpha}_{2,K-1}}{2} & \cdots & A^{\alpha}_{K-1,K-1} \end{pmatrix} \tag{10.8} $$

and in the vector

$$\xi^{\alpha} = \begin{pmatrix} B_1^{\alpha} \\ B_2^{\alpha} \\ \vdots \\ B_{K-1}^{\alpha} \end{pmatrix}. \qquad (10.9)$$

The chemical potentials are given as the derivatives of the previous formula:

$$^{P}\mu^{\alpha}(\mathbf{c}) = \frac{\partial\, ^{P}f^{\alpha}(\mathbf{c})}{\partial \mathbf{c}} = 2\Xi^{\alpha}\mathbf{c} + \xi^{\alpha}. \qquad (10.10)$$

By inverting this formula, the phase concentrations can be derived for the multi-component case as

$$^{P}\mathbf{c}^{\alpha}(\mu) = \frac{1}{2}\left(\Xi^{\alpha}\right)^{-1}\left(\mu - \xi^{\alpha}\right) \qquad (10.11)$$

and their derivatives with respect to μ are

$$\frac{\partial\, ^{P}\mathbf{c}^{\alpha}(\mu)}{\partial \mu} = \frac{1}{2}\left(\Xi^{\alpha}\right)^{-1}. \qquad (10.12)$$

Finally, the grand potential density in matrix notation reads as

$$^{P}\Psi^{\alpha}(\mu) = {}^{P}f^{\alpha}\left({}^{P}\mathbf{c}^{\alpha}(\mu)\right) - \left\langle \mu, {}^{P}\mathbf{c}^{\alpha}(\mu)\right\rangle. \qquad (10.13)$$

10.2 Formulation for the ideal solution model

Another possibility to describe the thermodynamic properties of phases in the context of the grand potential model is to treat them as ideal solutions. To distinguish it from the ideal part of the CALPHAD method, an ideal solution is indicated in the following by a capital I in the pre-superscript. The specific

Gibbs energy formulation for binary ideal solutions is introduced in section 3.2.2. For the corresponding modeling of multi-component systems, the free energy densities can be described with the function

$$ {}^{I}f^{\alpha}(T,\mathbf{c},c_K) = \sum_{i=1}^{K} \left({}^{\circ}f_i^{\alpha}(T)c_i + \frac{RT}{V_m} c_i \ln(c_i) \right) \tag{10.14} $$

including one coefficient ${}^{\circ}f_i^{\alpha}(T)$ for each component. To get the thermodynamic functions needed for the grand potential formulation explained in chapter 7.2, one can derive the chemical potential according to equation 7.10 as

$$ \begin{aligned} {}^{I}\mu_i^{\alpha}(T,\mathbf{c},c_K) &= \frac{\partial\, {}^{I}f^{\alpha}(T,\mathbf{c},c_K)}{\partial c_i} - \frac{\partial\, {}^{I}f^{\alpha}(T,\mathbf{c},c_K)}{\partial c_K} \\ &= {}^{\circ}f_i^{\alpha}(T) - {}^{\circ}f_K^{\alpha}(T) + \frac{RT}{V_m} \ln\left(\frac{c_i}{c_K}\right) \end{aligned} \tag{10.15} $$

and from this function one can solve for the phase concentrations of the independent components, which results in the equation

$$ {}^{I}c_i^{\alpha}(T,\mu_i,c_K) = c_K \exp\left(\frac{V_m}{RT} \left(\mu_i - {}^{\circ}f_i^{\alpha}(T) + {}^{\circ}f_K^{\alpha}(T)\right) \right). \tag{10.16} $$

To replace the dependent concentration c_K, one can make use of the summation condition $c_K = 1 - \sum_{j=1}^{K-1} c_j$ and insert the previous equations for the phase concentrations into it, leading to

$$ {}^{I}c_K^{\alpha}(T,\boldsymbol{\mu}) = \frac{1}{1 + \sum\limits_{j=1}^{K-1} \exp\left(\frac{V_m}{RT} \left(\mu_j - {}^{\circ}f_j^{\alpha}(T) + {}^{\circ}f_K^{\alpha}(T)\right) \right)}. \tag{10.17} $$

The final expression for the phase concentrations in the case of an ideal solution with multiple components follows as

$$
{}^{I}c_i^{\alpha}(T,\boldsymbol{\mu}) = \frac{\exp\left(\frac{V_m}{RT}\left(\mu_i - {}^{\circ}f_i^{\alpha}(T) + {}^{\circ}f_K^{\alpha}(T)\right)\right)}{1 + \sum_{j=1}^{K} \exp\left(\frac{V_m}{RT}\left(\mu_j - {}^{\circ}f_j^{\alpha}(T) + {}^{\circ}f_K^{\alpha}(T)\right)\right)} \tag{10.18}
$$

and can be used to derive the grand potential densities as

$$
{}^{I}\Psi^{\alpha}(T,\boldsymbol{\mu}) = {}^{I}f^{\alpha}(T, {}^{I}\mathbf{c}^{\alpha}(T,\boldsymbol{\mu}), {}^{I}c_K^{\alpha}(T,\boldsymbol{\mu})) - \sum_{j=1}^{K-1} {}^{I}c_j^{\alpha}(T,\boldsymbol{\mu})\mu_j. \tag{10.19}
$$

and to obtain the derivatives $\frac{\partial\, {}^{I}c_i^{\alpha}(T,\boldsymbol{\mu})}{\partial \mu_j}$, which are needed in order to calculate the mobilities according to equation 7.24. An ideal solution formulation similar to the one used for the entropy model in [65, 43] is for example utilized to simulate dendritic solidification and fragmentation of the system Al-Cu with the grand potential model in [7, 66, 67].

11 Determination of model parameters from CALPHAD data

The usage of thermodynamic databases in phase-field simulations is about to become a common practice and different strategies have been worked out to realize the coupling for the various phase-field formulations in an efficient manner (as described in the introduction section III). Possible formulations for the thermodynamic functions of the grand potential model are introduced in the previous section and the present section addresses how their unknown coefficients can be identified from CALPHAD data. One approach to model the free energy densities in a simplified way is to use quadratic approximations and derive their polynomial coefficients by Taylor expansion. This method has already proven its capability in different phase-field studies with the grand potential model [7, 63, 68]. For further information the reader is referred to a detailed article from Choudhury et al. [56]. An alternative approach to derive the coefficients for quadratic approximations or for ideal solution models is given by the least squares method, which gets addressed in section 11.2.

11.1 Taylor expansion

Every smooth function can be expressed as an infinite Taylor series around a single point. An approximation of an N times differentiable function in the vicinity of a certain point can be derived by using only the first $N+1$ terms of the Taylor series and the resulting polynomial is called Taylor polynomial of N-th order. This method of approximation can also be used to derive simplified expressions for the thermodynamic quantities in the framework of the grand potential model.

11.1.1 Approximation of the coefficients of the polynomial model

For a molar Gibbs energy $G_m^\alpha(c)$, which is expressed as a function of only one independent composition, a polynomial expression can be derived by performing the second-order Taylor expansion around a composition c_0 and get the approximated free energy densities by dividing through the molar volume:

$$f_T^\alpha(c) = \frac{G_m^\alpha(c_0)}{V_m} + \frac{G_m^{\alpha\prime}(c_0)}{V_m}(c-c_0) + \frac{G_m^{\alpha\prime\prime}(c_0)}{2V_m}(c-c_0)^2. \tag{11.1}$$

This is equivalent to the polynomial equation 10.1 with the coefficients chosen as

$$A^\alpha = \frac{G_m^{\alpha\prime\prime}(c_0)}{2V_m}, \tag{11.2}$$

$$B^\alpha = \frac{G_m^{\alpha\prime}(c_0)}{V_m} - \frac{G_m^{\alpha\prime\prime}(c_0)}{V_m}c_0 \tag{11.3}$$

as well as

$$X^\alpha = \frac{G_m^\alpha(c_0)}{V_m} - \frac{G_m^{\alpha\prime}(c_0)}{V_m}c_0 + \frac{G_m^{\alpha\prime\prime}(c_0)}{2V_m}c_0^2. \tag{11.4}$$

If the required analytical derivatives $G_m^{\alpha\prime}(c_0)$ and $G_m^{\alpha\prime\prime}(c_0)$ are not known, their values can be estimated by numerical differentiation. A visualization of the present approach can be found in figure 11.1(a). The concentration c_0, around which the expansion is done, hereby has to be chosen in dependence of the simulative task. Usually the equilibrium concentrations for the relevant equilibria between the involved phases are taken, because the deviations from equilibrium are only small for simulations of low undercoolings. Following the procedure explained in section 10.1, the thermodynamic functions needed for the grand potential model can be derived as:

$$\mu_T^\alpha(c) = f_T^{\alpha\prime}(c) = \frac{G_m^{\alpha\prime}(c_0)}{V_m} + \frac{G_m^{\alpha\prime\prime}(c_0)}{V_m}(c - c_0) \tag{11.5}$$

$$c_T^\alpha(\mu) = c_0 + \frac{\mu V_m - G_m^{\alpha\prime}(c_0)}{G_m^{\alpha\prime\prime}(c_0)} \tag{11.6}$$

$$\frac{\partial c_T^\alpha(\mu)}{\partial \mu} = \frac{V_m}{G_m^{\alpha\prime\prime}(c_0)} \tag{11.7}$$

$$\begin{aligned}\Psi_T^\alpha(\mu) &= f_T^\alpha(c_T^\alpha(\mu)) - \mu c_T^\alpha(\mu) \\ &= \frac{G_m^\alpha(c_0)}{V_m} - \mu c_0 - \frac{V_m\left(\frac{G_m^{\alpha\prime}(c_0)}{V_m} - \mu\right)^2}{2G_m^{\alpha\prime\prime}(c_0)}.\end{aligned} \tag{11.8}$$

The method of expanding around a certain point in the space of temperature and composition can also be applied for multi-component systems with more than two components. Analogously to the binary case, a second order approximation can be derived for constant temperature around a certain composition defined by the vector $\mathbf{c}_0$. Using Taylor's theorem for multivariable functions leads to an expression

$$f_T^\alpha(\mathbf{c}) = f^\alpha(\mathbf{c}_0) + (\mathbf{c} - \mathbf{c}_0)^T \nabla f^\alpha(\mathbf{c}_0) + \frac{1}{2}(\mathbf{c} - \mathbf{c}_0)^T \mathbf{H}_{f^\alpha}(\mathbf{c}_0)(\mathbf{c} - \mathbf{c}_0) \tag{11.9}$$

with $(\mathbf{c} - \mathbf{c}_0)^T$ as the transposed vector of $(\mathbf{c} - \mathbf{c}_0)$. Furthermore, $\mathbf{H}_{f^\alpha}(\mathbf{c})$ denotes the Hessian matrix, which includes the second order partial derivatives of $f^\alpha(\mathbf{c}) = G_m^\alpha(\mathbf{c})/V_m$, whereby the Gibbs energy is expressed as a function of only the independent compositions. For the multi-component formula written in the matrix notation from equation 10.7, the coefficients can be identified as

$$\boldsymbol{\Xi}^{\alpha} = \frac{1}{2}\mathbf{H}_{f^{\alpha}}(\mathbf{c}_0) \tag{11.10}$$

$$\boldsymbol{\xi}^{\alpha} = \nabla f^{\alpha}\mathbf{c}_0 - \mathbf{H}_{f^{\alpha}}(\mathbf{c}_0)\mathbf{c}_0 \tag{11.11}$$

$$X^{\alpha} = f^{\alpha}(\mathbf{c}_0) - \mathbf{c}_0^T\nabla f^{\alpha}(\mathbf{c}_0) + \frac{1}{2}\mathbf{c}_0^T\mathbf{H}_{f^{\alpha}}(\mathbf{c}_0)\mathbf{c}_0. \tag{11.12}$$

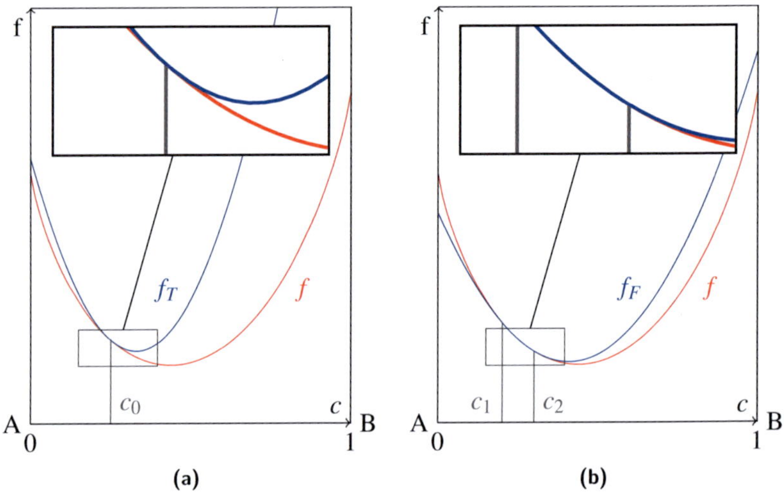

Figure 11.1: Illustration of two different methods to approximate the CALPHAD data: (a) shows the Taylor expansion $f_T(c)$ of the original function $f(c)$ around c_0, while (b) illustrates an approximation of $f(c)$ by the model function $f_F(c)$ fitted with the least squares method between c_1 and c_2.

If the polynomial is written in the form like in equation 10.6, the coefficients A_{ij}^{α} can be determined by matching the second derivatives of the polynomial free energies with the ones of the Gibbs energies from the database as

$$A^{\alpha}_{i,j} = \frac{1}{V_m} \frac{\partial^2 G^{\alpha}_m(\mathbf{c})}{\partial c_i \partial c_j}\bigg|_{\mathbf{c}_0} \quad \text{for } i < j \tag{11.13}$$

and also

$$A^{\alpha}_{i,i} = \frac{1}{2V_m} \frac{\partial^2 G^{\alpha}_m(\mathbf{c})}{\partial c_i^2}\bigg|_{\mathbf{c}_0}. \tag{11.14}$$

The coefficients B^{α}_i can then be determined as

$$B^{\alpha}_i = \left(\mu^{\alpha}_i(\mathbf{c}) - 2A^{\alpha}_{i,i} c_i - \sum_{\substack{j \\ j<i}}^{K-1} A^{\alpha}_{j,i} c_j - \sum_{\substack{j \\ j>i}}^{K-1} A^{\alpha}_{i,j} c_j \right)\bigg|_{\mathbf{c}_0} \tag{11.15}$$

and finally the coefficients X^{α} can be fixed by as

$$X^{\alpha} = \left(f^{\alpha}(\mathbf{c}) - \sum_{\substack{i=1,j=1 \\ i<=j}}^{K-1} A^{\alpha}_{i,j} c_i c_j - \sum_{i=1}^{K-1} B^{\alpha}_i c_i \right)\bigg|_{\mathbf{c}_0}. \tag{11.16}$$

A detailed explanation of the coupling method is given in [56, 48] and an example for such an approximation can be found in section 14.2.2.

11.1.2 Approximation of the coefficients of the ideal solution model

If the free energy densities are approximated with an ideal solution model like in equation 10.14, the coefficients can again be derived by Taylor expansion. For this purpose it is useful to express the ideal solution approximation as a function of solely the independent concentrations

$$
\begin{aligned}
{}^{I}f^{\alpha}(T,\mathbf{c}) = \sum_{i=1}^{K-1} \left({}^{\circ}f_i^{\alpha}(T)c_i + \frac{RT}{V_m} c_i \ln(c_i) \right) + \\
\left(1 - \sum_{i=1}^{K-1} c_i \right) \left({}^{\circ}f_K^{\alpha}(T) + \frac{RT}{V_m} \ln \left(1 - \sum_{i=1}^{K-1} c_i \right) \right).
\end{aligned}
\tag{11.17}
$$

The terms from configurational entropy, which include the logarithmic functions, are identical for the approximated formulation and the original one and therefore do not have to be taken into account to identify the unknown coefficients. To approximate only the other terms of the original formulation, a reduced function for the free energy densities from CALPHAD can be defined as

$$
\begin{aligned}
{}^{red}f^{\alpha}(T,\mathbf{c}) = \frac{G_m^{\alpha}(T,\mathbf{c})}{V_m} - \frac{RT}{V_m} \left(\sum_{i=1}^{K-1} c_i \ln(c_i) \right) - \\
\frac{RT}{V_m} \left(1 - \sum_{i=1}^{K-1} c_i \right) \ln \left(1 - \sum_{i=1}^{K-1} c_i \right).
\end{aligned}
\tag{11.18}
$$

By performing a first-order Taylor expansion of these reduced free energy densities and adding the configurational terms again, the ideal solution approximation follows as

$$
\begin{aligned}
{}^{I}f^{\alpha}(T,\mathbf{c}) = {}^{red}f^{\alpha}(T,\mathbf{c}_0) + \left(\sum_{i=1}^{K-1} (c_i - c_{0,i}) \frac{\partial\, {}^{red}f^{\alpha}(T,\mathbf{c}_0)}{\partial c_i} \right) + \\
\frac{RT}{V_m} \left(\sum_{i=1}^{K-1} c_i \ln(c_i) \right) + \frac{RT}{V_m} \left(1 - \sum_{i=1}^{K-1} c_i \right) \ln \left(1 - \sum_{i=1}^{K-1} c_i \right),
\end{aligned}
\tag{11.19}
$$

which is accurate for $\mathbf{c}_0$ and also has the same chemical potentials like the original function at this composition.

11.2 Approximating the thermodynamic functions with the least squares method

As an alternative to the previously described approach for deriving the unknown coefficients, one can make use of the methods from the field of regression analysis, which are used to estimate the dependency of a variable on one or more independent variables. These methods are often applied to establish a relationship between observable quantities for being able to make predictions based on measurements of the independent variables. In the context of quantitative phase-field modeling, the approach of data fitting can be used to derive simplified thermodynamic functions from the CALPHAD data with an acceptably small error inside of the data range of interest. Instead of expanding around one composition like it is the case for Taylor approximations, data from a range of compositions is used for the curve fitting, as illustrated in figure 11.1(b). In the scope of this thesis, the least squares method is applied to perform the fits, which is a commonly used tool for regression analysis. Hereby the coefficients of the chosen formulation are calculated such, that the sum of the squared differences between the values of the original data points and their approximation is minimized. This produces solutions, which do not have to be exact at a specific point, but lead to a small overall deviation inside of the whole range, from which the data points are taken. The fitting can be done for the previously introduced formulations, such that all of the required functions result from the determination of the free energy density coefficients. This procedure gets explained at the example of Fe-Cu in section 15 and at the example of an idealized system in section 12.2.2. Instead of using functions of concentration, the formulations can also be approximated in dependence of the chemical potentials like in section 11.4. Alternatively the model functions can be chosen independently from each other and the parameters can be derived by separate fittings, as it is explained in the following.

The method of least squares can be used for multi-component systems, as it is in principal suitable for any dimension. In this thesis, it is however only applied for binaries and gets described for such systems in the following. At first, an approach to derive expressions for the phase concentrations as functions of the chemical potential is explained. A prerequisite for this procedure is the invertibility of the chemical potential functions, which is only given if the chemical potentials are either strictly increasing or decreasing with concentration. If this requirement is fulfilled, one can calculate data points of $\mu^{\alpha}(c) = \frac{1}{V_m} \frac{\partial G_m^{\alpha}(c)}{\partial c}$ for different c

inside of the concentration regime of interest and at the fixed temperature. By exchanging the abscissa and the ordinate in each ordered pair, one can then derive data points for c over μ^{α}. An arbitrary model function, for example a polynomial, can then be fitted through these points to get $c^{\alpha}(\mu)$. The function for the partial derivative of the phase concentration with respect to the chemical potential does not have to be fitted separately but can be derived by analytical differentiation of the model function. If one has for example chosen a polynomial of degree n for $c^{\alpha}(\mu)$, then one can get a polynomial of degree $n-1$ for $\partial c^{\alpha}(\mu)/\partial \mu$. To approximate the grand potential densities $\Psi^{\alpha} = f^{\alpha}(\mu) - \mu c$, one can use the previously calculated ordered pairs of c^{α} and μ^{α} and together with the Gibbs energies from the database one can obtain ordered pairs of μ^{α} and the grand potential densities for the same concentration range as in the previous fits. By choosing for instance a polynomial approach, a function can be derived as a fit over the chemical potentials, like it is done for the system Al-Si in section 14.1. If an ideal solution formulation is chosen to describe the free energy densities, then the unknown parameters can be determined by least squares fitting in a similar way as described in section 11.1.2 by fitting only the reduced functions defined by equation 11.18.

11.3 Models for the temperature dependence

Phase-field simulations of alloy solidification are often carried out on the assumption of isothermal conditions and also most of the results presented in this thesis are performed for constant temperature. Therefore, the previously introduced procedures for the utilization of CALPHAD data only cover cases of fixed temperature. In this paragraph, the modeling of temperature dependent thermodynamic functions for the grand potential model is discussed without going into great detail. In principal the temperature is just an additional dimension and the approaches for multi-component systems can also be applied to account for the influence of temperature variations. The modeling can thus be realized by performing Taylor expansions for multiple variables (including the temperature) or by fitting temperature dependent functions with the least squares method using data of different compositions and temperatures. In this section, another approach is described, which is also applied for the modeling of iron-carbon in section 11.4.

The basic idea of the approach is to apply the previously described methods for two different temperatures and to combine the resulting expressions into temperature dependent functions by linear interpolation. In doing so, the two temperatures can be chosen as the lower and upper temperature bound of the simulation if the resulting range is small enough, such that the linear temperature dependence is an acceptable approximation. For example a thermodynamic function of concentration $f_0(\mathbf{c})$ can be fitted by a polynomial at the lower temperature T_0 and a second function $f_1(\mathbf{c})$ can be determined in the same way and for the same composition range at the higher temperature T_1. The temperature dependence of the thermodynamic function $f(T,\mathbf{c})$ inside of the chosen range can then be modeled by interpolating linearly between the two polynomials:

$$f(T,\mathbf{c}) = f_0(\mathbf{c}) + \frac{(f_1(\mathbf{c}) - f_0(\mathbf{c}))(T - T_0)}{T_1 - T_0}. \tag{11.20}$$

Thermodynamic functions being dependent on chemical potentials instead of concentrations can be modeled in the same way (like the grand potential densities in equation 11.21). Instead of fitting with the least squares method, the isothermal functions at the two temperatures can also be approximated by second order Taylor expansions with respect to the concentrations (as described in section 11.1). For the two temperatures and at the chosen compositions, around which the expansions are performed, such approximations reproduce the value of the function and its first and second order partial derivatives with respect to the concentrations. The partial derivatives with respect to temperature are not reproduced exactly with this method. In contrast, the value of the function and all of the first and second order partial derivatives at the specific composition and temperature are reproduced for a second order Taylor approximation in the space of composition and also temperature.

The assumption of a linear temperature dependence can be justified by analyzing the original Gibbs energies from the thermodynamic databases. In the CALPHAD formulation, the influence of temperature on the total Gibbs energies (equation 5.1) depends on the modeling of the individual terms. First of all, the ideal part ${}^{id}G_m^{\alpha}$ is linear in temperature for any material system as defined by equation 5.3. The temperature dependence of the reference part is determined by the Gibbs energies of the constituents of the specific phase. For the developers of thermodynamic datasets it is customary to model them as piecewise functions of temperature, consisting of power series with a preferably small number of coefficients, like in equation 5.16. Most of the interaction parameters used in the excess part of the CALPHAD formulation are chosen to be linear in temperature, as described in section 5.1.6. Because the ideal part is always linear in temperature and the interaction parameters from the excess part share this characteristic in most cases, the total Gibbs energies, chemical potentials or other thermodynamic properties also show a nearly linear behavior inside of a small temperature interval. Under the condition of only small temperature variations, it is thus justified to model the thermodynamic functions for the grand potential model to be linear in temperature. This is also consistent with the additional part of the evolution equations for the chemical potentials in equation 7.28, which only considers the first partial derivative of the phase concentrations with respect to temperature $\partial c_i^\alpha(T,\boldsymbol{\mu})/\partial T$. For the range of temperatures examined by simulations in the later chapters, all of the functions perform a nearly linear behavior. If a linear formulation is not sufficient to describe the temperature dependence of the thermodynamic quantities, a formulation of higher order can be chosen for the interpolation between the isothermal solutions or the approximation can be done by second order Taylor expansions in the space of composition and temperature.

As an alternative to the previously described approach, the thermodynamic parameters of a model with a linear temperature dependence can also be identified from the equilibrium compositions and slopes of the solidus and liquidus lines at a specific temperature, like for peritectic Ni-Al [69]. Furthermore, examples of temperature dependent dendritic and eutectic growth modeled with an ideal solution formulation can be found in [7] and [70], respectively. A detailed explanation about the modeling of linearly temperature dependent functions for multi-component systems is given in Choudhury et al. [56].

11.4 Modeling approach for stoichiometric phases

The formulation of stoichiometric phases in the CALPHAD method gets addressed in section 5.1.7. In the following, the modeling of a stoichiometric phase in the thermodynamic framework of the grand potential model gets explained at the example of cementite. This coupling approach for stoichiometric phases has already been applied in a phase-field study about the influence of diffusivity on the eutectoid transformation in Fe-C [63] and the following investigation on deviations from the cooperative growth mode [71]. Both simulation studies thematize the eutectoid transformation, during which the austenite phase (denoted as γ) gets replaced by the lamellar pearlite structure. As the pearlite is composed of the iron-rich ferrite phase (denoted as α) and the stoichiometric cementite phase Fe_3C (denoted as β), in total three different phases are involved in the transformation. The part of the phase-diagram around the eutectoid point is shown in figure 11.2. This diagram is calculated from an assessment of Gustafson et al. [72] and all thermodynamic parameters of the phase-field studies are derived from this database. The Gibbs energy descriptions from the CALPHAD dataset include magnetic contributions (as described in section 5.1.8) for both the ferrite and the austenite phase. As the approximations are done for the total Gibbs energies of the phases, these contributions are included in the simplified formulations without being treated separately. The magnetic contributions are however not the subject of discussion of this section, but the modeling of the stoichiometric cementite phase.

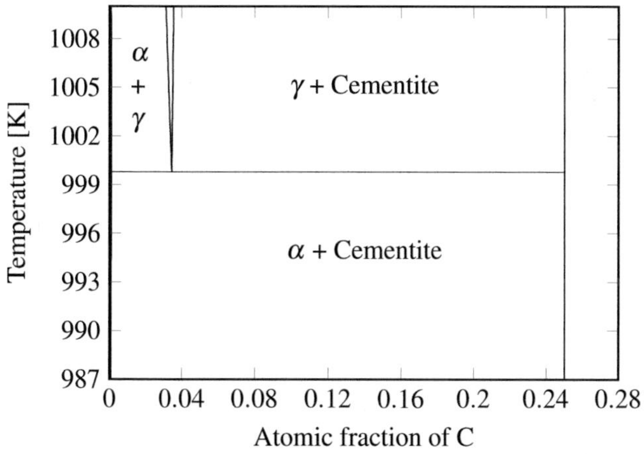

Figure 11.2: Part of the phase-diagram of iron-carbon around the eutectoid point, which is calculated from the original CALPHAD dataset.

In order to keep the important characteristics of the original phase diagram for an approximated formulation of the thermodynamic parameters, the relevant equilibria between the different phases have to be reproduced. According to the stoichiometry, the Gibbs energy of the Fe_3C phase is defined in the database as a single point with an atomic fraction of carbon of 0.25 (a definition similar to the one depicted in figure 5.1(a)). For the simulation with the phase-field method, cementite is modeled as a composition dependent phase by describing its Gibbs energy curve as a steep parabola (a definition corresponding to figure 5.1(b)). Such a modeling of line compounds is common in the context of the phase-field method and gets discussed in Hu et al. [73]. As already addressed in section 10.1, the quadratic description of the Gibbs energies as functions of concentration is equivalent to applying polynomials of second degree in the chemical potential for the grand potentials. By choosing the chemical potential of carbon as the independent chemical potential μ (similarly c denotes the concentration of carbon in this section), the grand potential densities of all three phases are modeled as functions of the form:

$$\Psi^{\alpha,\beta,\gamma}(T,\mu) = A^{\alpha,\beta,\gamma}(T)\mu^2 + B^{\alpha,\beta,\gamma}(T)\mu + C^{\alpha,\beta,\gamma}(T). \tag{11.21}$$

Applying the method described in section 11.3, the coefficients for all three phases are determined by an interpolation over temperature from their values at the eutectoid temperature T_e and at a temperature T_1, which is 10 K below T_e. A linear temperature dependence is assigned to the phases by choosing their coefficients as:

$$A^{\alpha,\beta,\gamma}(T) = A^{\alpha,\beta,\gamma}_{T_1} + (T - T_1)\frac{A^{\alpha,\beta,\gamma}_{T_e} - A^{\alpha,\beta,\gamma}_{T_1}}{T_e - T_1} \tag{11.22}$$

(the same applies to $B^{\alpha,\beta,\gamma}(T)$ and $C^{\alpha,\beta,\gamma}(T)$).

The coefficients at the eutectoid temperature are determined by using the characteristics of a three phase equilibrium. If all the phases are at their specific equilibrium compositions, they have the same chemical potential μ_{eq}, as well as they share the same grand potential Ψ_{eq}. This means, that a common tangent can be applied to their Gibbs energies, as schematically shown in figure 11.3(a). To reproduce this property with the approximated formulations, the coefficients are determined for the equilibrium compositions $c^{\alpha,\beta,\gamma}_{eq}$. The first coefficients $A^{\alpha,\gamma}_{T_e}$ of the austenite and ferrite phase thus are calculated from the second derivatives of the Gibbs energies as:

$$A_{T_e}^{\alpha,\gamma} = \frac{1}{2}\frac{\partial^2 \Psi^{\alpha,\gamma}(T,\mu)}{\partial \mu^2}\bigg|_{T_e,\mu_{eq}} = \frac{\partial c^{\alpha,\gamma}(T,\mu)}{\partial \mu}\bigg|_{T_e,\mu_{eq}} = \frac{-V_m}{\frac{\partial^2 G_m^{\alpha,\gamma}(T,c)}{\partial c^2}\Big|_{T_e,c_{eq}^{\alpha,\gamma}}}. \tag{11.23}$$

Because cementite is modeled as a single point in the CALPHAD dataset and therefore the second derivative of its Gibbs energy with respect to concentration cannot be calculated, this procedure is not applicable for cementite. Instead, a high curvature is assigned to its parabolic Gibbs energy curve by choosing a small value of $-1.9 \times 10^{-14}\mathrm{m}^3/\mathrm{J}$ for $A_{T_e}^{\beta}$. Hereby, the coefficient is arbitrarily chosen and not much relevance should be attributed to its absolute value. The main point is, that such a choice for $A_{T_e}^{\beta}$ corresponds to a small value of $\partial c^{\beta}/\partial\mu$, which ensures (as already discussed at the end of section 12.2.1) only small deviations from the stoichiometric composition of cementite during the simulation. For all three phases the parameters $B_{T_e}^{\alpha,\beta,\gamma}$ are fixed as

$$B_{T_e}^{\alpha,\beta,\gamma} = -c_{eq}^{\alpha,\beta,\gamma} - 2A_{T_e}^{\alpha,\beta,\gamma}\mu_{eq} = -c_{eq}^{\alpha,\beta,\gamma} - \frac{2A_{T_e}^{\alpha,\beta,\gamma}}{V_m}\frac{\partial G_m^{\alpha,\gamma}(T,c)}{\partial c}\bigg|_{T_e,c_{eq}^{\alpha,\gamma}}, \tag{11.24}$$

whereby the derivatives of the Gibbs energies are given for the ferrite and the austenite phase. The derivative for cementite cannot be derived, but due to the condition of equal chemical potentials the values from the other two phases can also be used for the parameter identification of this stoichiometric phase. The third parameters $C_{T_e}^{\alpha,\beta,\gamma}$ of all phases are fixed as

$$C_{T_e}^{\alpha,\beta,\gamma} = \frac{G_m^{\alpha,\beta,\gamma}(T_e, c_{eq}^{\alpha,\beta,\gamma})}{V_m} + A_{T_e}^{\alpha,\beta,\gamma}\mu_{eq}^2 \tag{11.25}$$

to reproduce Ψ_{eq} if the values of the eutectoid equilibrium are inserted into formula 11.21.

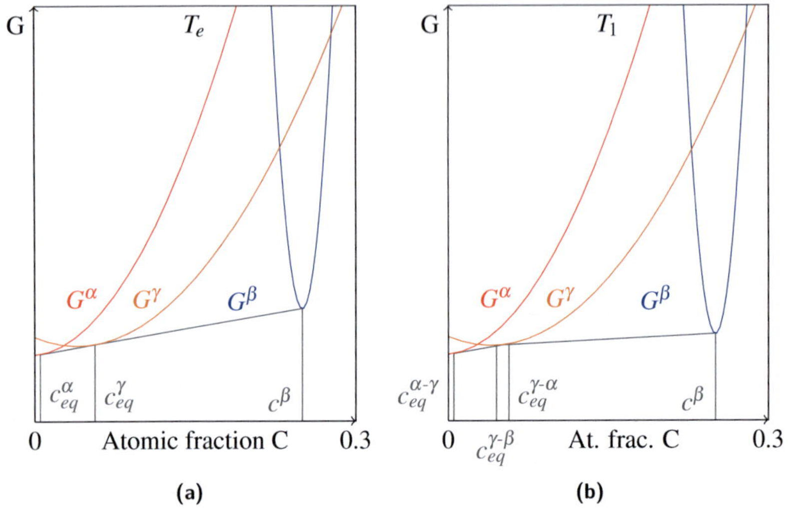

Figure 11.3: Schematic illustration of the approximated Gibbs energies and their common tangents at (a) the eutectoid temperature T_e and (b) a temperature $T_1 < T_e$. The Gibbs energy of the cementite phase β is given as a single point in the CALPHAD dataset. In these diagrams G^β is represented by a steep parabola, as it is modeled for the phase-field simulation. At the eutectoid temperature all three phases are in equilibrium, whereas at the lower temperature one equilibrium between the austenite phase γ and the ferrite phase α exists and a second equilibrium is given between austenite and cementite. (The common tangent for the third equilibrium between ferrite and cementite is not shown.)

Not only one, but three different equilibria can be found between the phases at the lower temperature, which has to be considered in the determination of the thermodynamic parameters for T_1. However, not every equilibrium can be exactly reproduced with the given number of coefficients and therefore the approximation is done in the following manner. The coefficients for the ferrite and austenite phase are derived in the same way as for the eutectoid temperature. Because the phases have no unique equilibrium composition at the lower temperature, their eutectoid compositions are chosen for the calculation of the coefficients. Again a high curvature is assigned to the Gibbs energy of the cementite phase at the low temperature with $A_{T_1}^{\beta} = A_{T_e}^{\beta}$. The remaining coefficients are fixed in order to reproduce the equilibrium between cementite and austenite. This choice is made because the cementite forms from austenite and therefore it is of major importance to reproduce the equilibrium between these two phases correctly. For a better understanding how $B^{\beta}_{T_1}$ is fixed, the Gibbs energies and common tangents for the lower temperature T_1 are visualized schematically in figure 11.3(b). It can be seen, that the equilibrium concentration of the austenite phase $c^{\gamma\text{-}\alpha}_{eq}$ for the equilibrium with the ferrite phase differs from the concentration $c^{\gamma\text{-}\beta}_{eq}$ for the equilibrium with cementite. The chemical potential $\mu^{\gamma\text{-}\beta}_{eq} = \dfrac{G^{\gamma}_{m}(T_1, c^{\gamma\text{-}\beta}_{eq}) - G^{\beta}_{m}(T_1)}{(c^{\gamma\text{-}\beta}_{eq} - c^{\beta})V_m}$ for an equilibrium between the austenite and cementite phase is used for determining the second coefficient of the stoichiometric phase as

$$B^{\beta}_{T_1} = -c^{\beta} - 2A^{\beta}_{T_1} \cdot \mu^{\gamma\text{-}\beta}_{eq}. \tag{11.26}$$

Finally, the last coefficient for cementite is chosen as

$$C^{\beta}_{T_1} = \frac{G^{\beta}_{m}(T_1)}{V_m} + A^{\beta}_{T_1} \cdot \mu^{\gamma\text{-}\beta^{2}}_{eq} \tag{11.27}$$

to assign a grand potential of $\dfrac{G^{\beta}_{m}(T_1)}{V_m} - \mu^{\gamma\text{-}\beta}_{eq} c^{\beta}$ to the cementite phase.

Figure 11.4 shows the coexistence lines for the α and γ phase and the lines for the coexistence of γ and cementite, which are decisive for the eutectoid transformation. This diagram is recalculated from the approximated formulations and includes the cementite phase as a vertical line at $c = 0.25$. The two lines belonging to the equilibrium compositions of austenite intersect at the eutectoid point from the original diagram in figure 11.2 and the equilibrium composition of the ferrite phase is also well reproduced. The approach of treating a stoichiometric phase as a phase with a small range of solubility is thus appropriate to keep the features of the original phase diagram, which are important for the eutectoid transformation.

Even though the composition of a stoichiometric phase nearly does not change during the simulation, the evolution equations have to be solved for such a phase in the same manner as for the other phases. In fact, with constant compositions of the stoichiometric phases, the simulations could be performed in an optimized way. For future studies of stoichiometric phases, the calculation of the diffusion equation for such phases could be skipped, such that they would keep the same composition during the whole simulation as initially assigned to them.

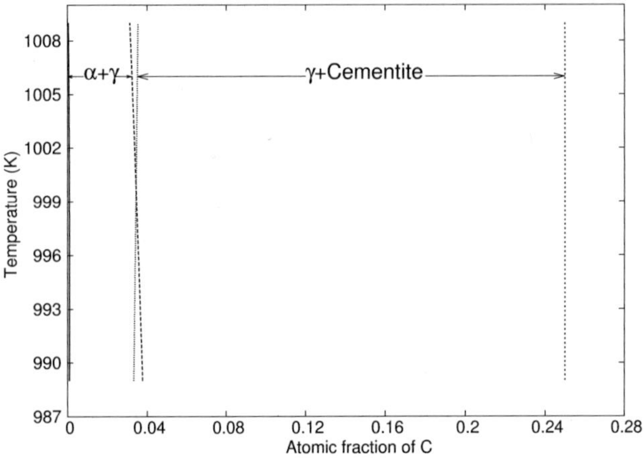

Figure 11.4: Part of the phase-diagram of Fe-C at the Fe-rich side around the eutectoid temperature. The diagram is recalculated from the approximated Gibbs energies. (Diagram from [63] with small modifications.)

11.5 Modeling pseudobinary phases

In section 3.6, the differences between pseudobinary and quasibinary subsystems are discussed. An example for the thermodynamic description of a subsystem from an industrially relevant alloy gets described in this paragraph. As this topic is not the main subject matter of this thesis, it is only addressed briefly. A quasibinary system is reported for the system Al-Cr-Ni [74, 75] and a diagram of this subsystem can be found in the book "Ternary alloys" from Rogl [76]. According to this diagram, the quasibinary system ranges from pure chromium to the intermetallic compound NiAl, which forms for the same amounts of nickel and aluminum. A congruent melting point $T_m = 1911$ K is reported for this composition, which constitutes the maximal melting temperature of the binary system Al-Ni. Because of this property, alloys based on the compound NiAl are well suited for high-temperature applications and feature excellent physical and mechanical properties if a third component like chromium is added.

In more recent publications about the system Al-Cr-Ni [77, 78] it is however mentioned, that it does not contain a quasibinary system. The hypothesis of a quasibinary system is also not supported by a thermodynamic dataset for the ternary system published by Dupin [79]. From this dataset a phase diagram can be calculated, which ranges from pure Cr to the congruent melting point, having a slightly higher aluminum content ($x_{Al} = 0.505$) than the stoichiometric composition assumed in the older publications. The diagram includes two solid phases, which both have a body-centered cubic lattice and differ by the occupation of the lattice sites. Hereby the Cr-rich phase BCC-A2 corresponds to the disordered state, whereas the ordered state is given by the intermetallic BCC-B2 phase. In contrast to the system reported in the book of Rogl [76], the calculated diagram does not contain a distinct eutectic point, but a region for which the liquid and the two BCC phases are in equilibrium. This three-phase region is an evidence for the non-existence of a quasibinary subsystem NiAl-Cr.

Due to their importance for high-temperature applications, it is of great interest to gain more knowledge about the formation of Al-Cr-Ni alloys by quantitative phase-field simulations based on CALPHAD data. However, the formulations used by Dupin [79] are too complex for the application in phase-field simulations. In her assessment, Dupin uses sophisticated Gibbs energy formulations to model the ordered and disordered states of the BCC crystals. By the use of several sublattices, the ordered phase is described together with the related disordered phase by only one single Gibbs energy function. This modeling approach is described in Kusoffsky et al. [80] for the case of face-centered cubic crystals. The applied method requires to solve for the site-fractions of the additional sublattices by Gibbs energy minimization, which in consequence increases the required computational effort compared to single-lattice models. The use of simplified expressions for the thermodynamic quantities is thus required to keep the computational effort within reasonable bounds. Here again the concept of quasibinaries comes into play, because a substantial simplification can be achieved, if the number of components is reduced by approximating the ternary system with the artificial binary system NiAl-Cr. A concept for such a treatment can be found in the paper of Choudhury et al. [56]. To approximate the free energy densities of the relevant phases at a fixed temperature, a quadratic formulation (see also section 10.1) is chosen, which only depends on the composition of one component. The temperature and composition for determining the polynomial coefficients is located inside the three-phase region of liquid and the BCC phases. From these Gibbs energies, the equilibrium concentrations for the pseudobinary are determined and used to fix the values of the coefficients. A detailed comparison between the binary approximation and a ternary modeling is however still missing and could be a topic for further research in this field.

12 The trade-off between accuracy and complexity

Before proceeding with the actual case studies for the utilization of thermodynamic databases, the previously introduced coupling approaches are examined theoretically on the basis of idealized systems in the following. The formulation of Gibbs energies for the calculation of phase diagrams involves a trade-off between the accuracy of the fit and the complexity of the description. These two aspects are also the main issues for choosing a formulation to describe the free energy densities and related thermodynamic quantities for the purpose of phase-field simulations. The trade-off is however different in this case, as the formulations only have to cover a small part of the phase diagram but influence the calculation effort needed for every computational cell and time step of large-scale and long-time simulations. In this section, both issues are addressed separately by comparison of the CALPHAD formulations and the different approximation approaches.

12.1 Computational effort

To quantify the computational effort of the original CALPHAD formulations and the approximations, the mathematical operations required for their calculation are analyzed in the following. Hereby, only binary systems at constant temperature are considered, for which the functions are dependent on one variable of concentration. Regarding the temperature dependence, it is computationally advantageous to treat problems as isothermal or to use a linear dependency if a variation of temperature has to be taken into account (as addressed in section 11.3). The calculation of the free energy density for the case of an isothermal quadratic approximation (equation 11.1) can be done according to Horner's method. This evaluation includes two additions and two multiplications of the coefficients given as floating-point numbers. The calculation of the ideal part of the CALPHAD formulation ${}^{id}G(c)$ given in equation 5.6 includes one addition and one subtraction, if the value of $1-c$ is calculated only once and cached for reuse. Furthermore two multiplications are required, what results in the same number of basic arithmetic operations as for the quadratic approximation so far. Finally, the required computational effort of the ideal part is significantly raised by the two calculations of the natural logarithm. There are different algorithms to evaluate the natural logarithm. For example, the calculation can include the computation of the arithmetic-geometric mean [81], which has to be solved for iteratively. In dependence of the required number of iterations, several basic arithmetic operations have to be performed for this evaluation. Therefore the calculation of the ideal part by itself entails a bigger computational effort than the quadratic scheme and the effort is further increased by the other terms of the CALPHAD formulation. For instance, the calculation of the reference part requires one additional addition and multiplication and moreover the excess part – depending on its order – requires several additions and multiplications.

Instead of evaluating them from mathematical expressions, the values for the thermodynamic quantities can also be received from look-up tables. These tables contain the values for the required quantities in dependence of the variables. The application of lookup-tables can thus minimize the computational effort at the expense of additional memory accessing. This strategy is not considered within the scope of this thesis but should be taken into account as a possible optimization for future phase-field studies.

12.2 Analysis of the deviation arising from a quadratic fit

As pointed out before, the approach of modeling the CALPHAD Gibbs energies by quadratic approximations is computationally advantageous, but also its accuracy has to be taken into account. The deviation of the resulting expressions from the original ones naturally depends on the complexity of the CALPHAD formulation. Moreover, the accuracy of the simplified model is also related to the composition, around which the approximation is performed. In order to study the quality of the quadratic approach, the individual parts of the CALPHAD formulation get analyzed separately in the following. First of all, the linear reference part ${}^{ref}G_m^\alpha$ (eq. 5.5) can be perfectly represented by a polynomial of second degree for all possible compositions. Secondly the excess part of the

Gibbs energies ${}^{ex}G_m^\alpha$ is given by Redlich-Kister formulations of different order k in equation 5.7. For a Redlich-Kister formulation of zeroth order, the excess part is quadratic in c and can thus be described with a polynomial of degree two. Excess parts of order $k = 1$ or higher can not be perfectly represented with a quadratic function. Therefore, the accuracy of the fits depends on the order of the excess part. Thirdly for the ideal part, a relation between the composition, around which the approximation is performed, and the quality of the approximation can be found and is discussed in the next paragraph.

12.2.1 Deviations due to Taylor approximations of the ideal part

In the following, the quadratic approximation of the ideal part of the CALPHAD formulation is examined in more detail. For the sake of simplicity, the approximation is done for a nondimensionalized version of the Gibbs energies and the subscript indicating molar values is not written. The nondimensionalized version of the ideal part results from equation 5.6 by dividing through R and the temperature:

$$
{}^{id}G(c) = \begin{cases} 0 & \text{if } c = 0 \text{ or } c = 1 \\ c\ln(c) + (1-c)\ln(1-c) & \text{if } 0 < c < 1 \end{cases} \tag{12.1}
$$

The domain of ${}^{id}G(c)$ is $[0,1]$ and it is differentiable inside of the open interval $(0,1)$. The first, second and third derivatives of ${}^{id}G(c)$ are

$$
{}^{id}G'(c) = \ln\left(\frac{c}{1-c}\right) = 2\,\text{artanh}(2c-1) \tag{12.2}
$$

$$
{}^{id}G''(c) = \frac{1}{1-c} + \frac{1}{c} \tag{12.3}
$$

$$ {}^{id}G'''(c) = \frac{1}{(1-c)^2} - \frac{1}{c^2}. \qquad (12.4) $$

For the ideal part the second-order Taylor approximation $G_T(c)$ around $c_0 \in (0,1)$ can be derived by using equation 11.1. The approximations of the original function at several compositions are plotted in figure 12.1.

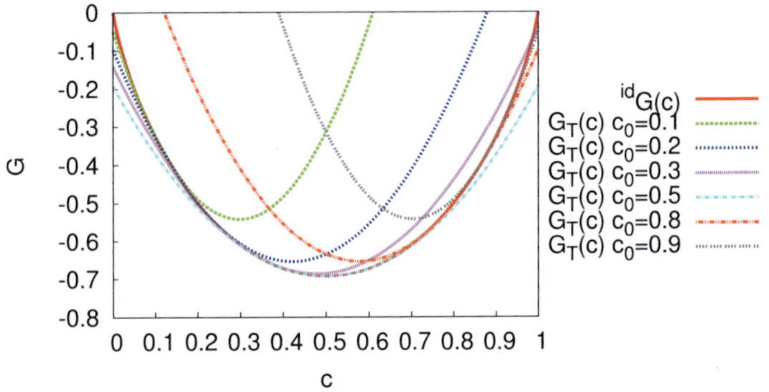

Figure 12.1: The original function ${}^{id}G(c)$ plotted together with the quadratic approximations $G_T(c)$ for different c_0. It can be seen, that if the curve for $c_0 = 0.8$ gets mirrored around the axis $c = 0.5$, it coincides with the curve for $c_0 = 0.2$.

As can be seen in the plots, ${}^{id}G(c)$ has an axis of symmetry $c = 0.5$ and each of the graphs is symmetric with regard to the axis $c = c_0$. Due to its symmetry, ${}^{id}G(c)$ is analyzed separately in the intervals $I_0 := [0, 0.5]$ and $I_1 := [0.5, 1]$, which both include $c = 0.5$. In the same way, the interval where the ideal part is differentiable, is divided into the half-closed intervals $\overline{I_0} := (0, 0.5]$ and $\overline{I_1} := [0.5, 1)$. In the diagram it can be seen, that for $c_0 \in \overline{I_0}$ the function value of $G_T(c)$ is smaller than the one of the original function in the range from zero to c_0, where the graphs intersect. The curve of the approximation is located above the original one for values bigger than $c_0 \in \overline{I_0}$. The plots also reveal, that there can be a second intersection point at $c_1 \neq c_0$, for example in the case $c_0 = 0.3$. For $c_0 = 0.5$ the quadratic fit is symmetric to the axis $c = 0.5$ and has smaller values than the original function for every $c \neq 0.5$. The curves of the approximations for $c_0 \in \overline{I_1}$ are the mirrored curves for $1 - c_0$ with respect to the axis $c = 0.5$.

The same properties can also be seen in the plots of the differences between the original function and the approximations

$$\Delta G(c) = {}^{id}G(c) - G_T(c) \tag{12.5}$$

for varying c_0 given in figure 12.2. If $\Delta G_{c_0}(c)$ denotes the difference for a Taylor approximation around c_0, then it can be used to express the difference $\Delta G_{(1-c_0)}(c)$ for an approximation around $1-c_0$ by $\Delta G_{c_0}(c) = \Delta G_{(1-c_0)}(1-c)$.

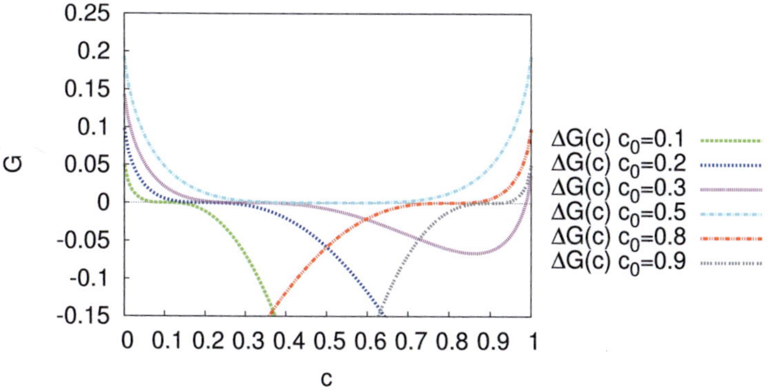

Figure 12.2: The differences $\Delta G(c)$ between ${}^{id}G(c)$ and the quadratic approximations $G_T(c)$ given as Taylor approximations around different compositions c_0.

Due to the symmetry of the problem it is thus sufficient to analyze the deviation of the fits only for $c_0 \in \overline{I_0}$. To analyze the accuracy of the fit, the absolute value of the difference $|\Delta G(c)|$ has to be examined. Inside of the interval I_0 the deviation $|\Delta G(c)|$ can be expressed as a piecewise function for any c_0 inside of $\overline{I_0}$. A separate discussion of this issue can be found in appendix A. It is justified to define the absolute value function inside of the interval I_0 as

$$|\Delta G(c)| = \begin{cases} \Delta G(c) & \text{if } c \leq c_0 \wedge c \in I_0, c_0 \in \overline{I_0} \\ -\Delta G(c) & \text{if } c > c_0 \wedge c \in I_0, c_0 \in \overline{I_0}. \end{cases} \tag{12.6}$$

To quantify the deviation of the approximation for values around c_0, the integral of $|\Delta G(c)|$ from $c_0-\varepsilon$ to $c_0+\varepsilon$ can be calculated. In the present context, ε denotes half of the width of the chosen concentration range and is not related to the interface width in the phase-field method. For $0 \le \varepsilon \le \min(c_0, 0.5-c_0)$ the integral can be expressed as

$$\int_{c_0-\varepsilon}^{c_0+\varepsilon} |\Delta G(c)|\, dc = \int_{c_0-\varepsilon}^{c_0} \Delta G(c)\, dc - \int_{c_0}^{c_0+\varepsilon} \Delta G(c)\, dc. \tag{12.7}$$

With the indefinite integral

$$\begin{aligned}\int \Delta G(c)\, dc =& \frac{1}{2}\Big(\tilde{G}(c) - c - {}^{id}G(c_0)\,(2c) - {}^{id}G'(c_0)\,(c^2 - 2c_0 c) - \\ & {}^{id}G''(c_0)\left(\frac{1}{3}c^3 - c^2 c_0 + c c_0^2\right)\Big) + C\end{aligned} \tag{12.8}$$

including the function

$$\tilde{G}(c) = c^2 \ln(c) - (1-c)^2 \ln(1-c) \tag{12.9}$$

the definite integral results as

$$\begin{aligned}\int_{c_0-\varepsilon}^{c_0+\varepsilon} |\Delta G(c)|\, dc =& \tilde{G}(c_0) - \frac{1}{2}\Big(\tilde{G}(c_0-\varepsilon) + \tilde{G}(c_0+\varepsilon)\Big) + \\ & {}^{id}G'(c_0)\varepsilon^2 - \frac{1}{3}{}^{id}G''(c_0)\left(c_0^2\varepsilon + c_0\varepsilon^2\right).\end{aligned} \tag{12.10}$$

For the special case $c_0 = 0.5$ the deviation $\Delta G_{c_0=0.5}(c)$ does not take negative values inside of the domain of definition, as it can be seen in figure 12.2 (see also appendix A). Therefore $|\Delta G_{c_0=0.5}(c)|$ can be equated with $\Delta G_{c_0=0.5}(c)$ and the definite integral for an integration around $c_0 = 0.5$ can be written as

$$\int_{0.5-\varepsilon}^{0.5+\varepsilon} |\Delta G(c)|\,dc = \frac{1}{2}\tilde{G}(0.5+\varepsilon) - \frac{1}{2}\tilde{G}(0.5-\varepsilon) - \varepsilon - 2\varepsilon \cdot {}^{id}G(0.5) - \frac{\varepsilon^3}{3}\,{}^{id}G''(0.5) \tag{12.11}$$

Figure 12.3(a) shows the graph of $\int_{c_0-\varepsilon}^{c_0+\varepsilon} |\Delta G(c)|dc/(2\varepsilon)$ for $\varepsilon = 0.05$. The function is plotted over c_0 ranging from ε to $0.5-\varepsilon$ and for the single value at $c_0 = 0.5$. Because $|\Delta G(c)|$ results from idealized and nondimensionalized Gibbs energies and the integration range is arbitrarily chosen, the absolute values, which are plotted, are not meaningful. However it gets obvious, that the function takes the biggest value for $c_0 = \varepsilon$ and with increasing c_0 the values of the function decrease and are minimal at $c_0 = 0.5$. It can thus be stated, at least for the binary case, that the graph of the ideal part from the CALPHAD Gibbs energy description ${}^{id}G(c)$ can be better resembled by a quadratic approximation, if it is done in the range of similar compositions of all components. Bigger deviations between the graph of the original function and the one of the Taylor series appear, if the approximation is performed for an alloy with one major component. An additional analysis concerning the deviations of the approximated phase concentration functions is documented in appendix B. As the functions of the phase concentrations are derived from the Gibbs energy functions, a similar dependence on c_0 can be found for their case.

These results should be interpreted cautiously, since the integration of the deviations is corresponding to the hypothetical case of uniformly distributed concentrations inside of the integration range around c_0. However, if the approximation is carried out adequately for the simulative task, compositions in the vicinity of c_0 occur more frequently during the simulation. In the PhD thesis [48, p. 176] of Abhik Choudhury, an equation is reported for the concentration deviation Δc^β of a phase β from its equilibrium concentration in contact with a phase α:

$$\Delta c^\beta = \frac{\sigma_{\alpha\beta}\,\kappa}{\frac{\partial^2 f^\beta}{\partial c^2}\left(c^\alpha - c^\beta\right)}. \tag{12.12}$$

This relation follows from equating the driving force for phase transformations in alloys to the capillary force, which enters the formula by the curvature κ multiplied with the surface tension $\sigma_{\alpha\beta}$ between the two phases. The driving force for phase transformations is given by the difference of the grand potentials and can be expressed by means of concentration differences and the second derivative of the free energy with respect to concentration, which leads to the above equation. The deviation analysis in this section only considers the free energies of single phases but does not take the relation between different phases or capillary effects into account and therefore no statements can be made on basis of this analysis about these aspects. Nevertheless, it gets clear from the above formula, that the deviations from the equilibrium concentration of a single phase are dependent on the second derivative of its free energy. This quantity, which is given for the idealized example by equation 12.3, enters the equation in the denominator and the reciprocal function $\left(\frac{\partial^2 f}{\partial c^2}\right)^{-1} = \frac{\partial c}{\partial \mu}$ are referred to as susceptibility in the following. A given gradient of the chemical potential thus results in a larger change of concentration if the value of the susceptibility is big. For the ideal case under consideration in the current analysis, this quantity is plotted over concentration in figure 12.3(b). In contrast to the graph in figure 12.3(a) the value of the susceptibility is maximal at $c = 0.5$ and vanishes for the pure components. These characteristics also manifest themselves in the plot of the free energies in figure 12.1, because a strong curvature of the parabola for $c_0 = 0.1$ corresponds to a small value of the susceptibility and a weak curvature for $c_0 = 0.5$ to a big value of the susceptibility. Taking this into account, the variation of concentrations close to $c = 0$ is much smaller, than for concentrations in the region around $c = 0.5$. The susceptibility thus acts compensating to the previously found dependence of the deviations of approximated Gibbs energies on c_0. In a region of composition, where a deviation from c_0 is connected to a bigger inaccuracy of the approximation, the occurring concentration variations are smaller. In conclusion, no definite statement can be made about the deviations due to quadratic formulations from the above derivation. As it follows from equation 12.12, the variation from the compositions, around which the expansions are made, is dependent on the interplay of capillary effects and the driving forces acting between the different phases and cannot be derived solely from the independent free energies of the phases. Statements about the quality of approximations should therefore be based on the criterion, if the important features of the individual simulative task can adequately be reproduced.

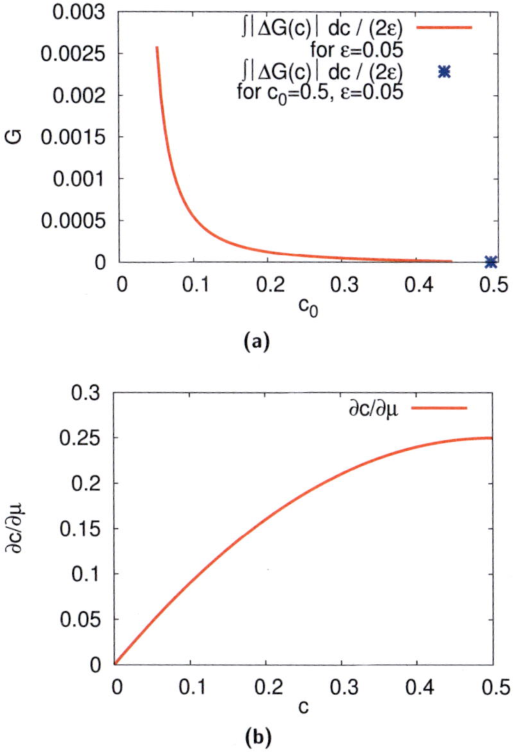

Figure 12.3: (a) The integral $\int_{c_0-\varepsilon}^{c_0+\varepsilon} |\Delta G(c)|\, dc/(2\varepsilon)$ is plotted for different Taylor approximations at compositions c_0. The plot shows the results for $\varepsilon = 0.05$. (b) The value of $\dfrac{\partial c}{\partial \mu}$ (derived from ${}^{id}G(c)$) is plotted over concentration.

12.2.2 Least squares fitting of the ideal part

For the comparison with the Taylor expansions, the ideal Gibbs energy defined by equation 12.1 is approximated again with quadratic polynomials $G_F(c) = Ac^2 + Bc + X$ by applying the least squares method (see also section 11.2). Several of these polynomial regressions are performed for sets of twenty equally distributed data points from different concentration ranges. To obtain comparable results, the concentration ranges are chosen as $c_0 - \varepsilon \leq c \leq c_0 + \varepsilon$ with the same value of $\varepsilon = 0.05$ as applied for the previous integrals of $|\Delta G(c)|$.

The averages of the absolute values of the differences between the original and the fitted values $\langle|\Delta G_F(c)|\rangle = \langle|^{id}G(c) - G_F(c)|\rangle$ are calculated for data ranges around different c_0. Hereby the average values are derived by summing up the absolute values of the differences between each value of the data points and its corresponding value from the fitted function and dividing the sum by the number of data points. The results are plotted in figure 12.4 together with the curve from figure 12.3(a).

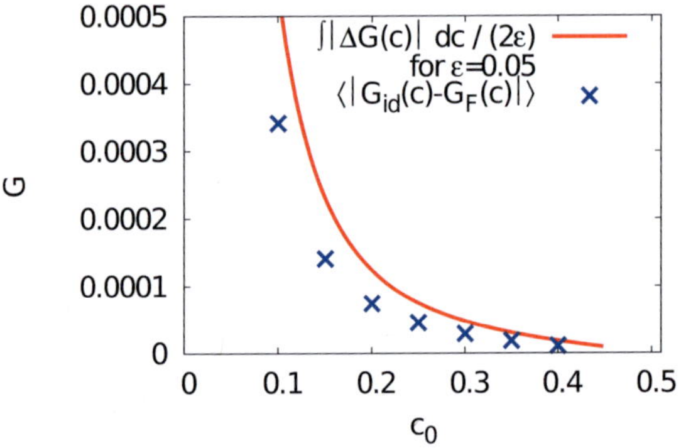

Figure 12.4: The curve for $\int_{c_0-\varepsilon}^{c_0+\varepsilon} |\Delta G(c)|\, dc/(2\varepsilon)$ from figure 12.3(a) is plotted together with the averages of the deviations $\langle|^{id}G(c) - G_F(c)|\rangle$ for quadratic fits with the least squares method inside of the composition ranges $c_0 - \varepsilon \leq c \leq c_0 + \varepsilon$ for $\varepsilon = 0.05$.

As it can be seen, the averaged deviations for the least squares fits are in the same way dependent on c_0 as the ones from the Taylor expansions. Furthermore, the averaged errors caused by polynomial regression are smaller for all of the evaluated c_0. This result is however no surprise, as the least squares method is designed to minimize the overall error of estimation for a range of data and not to reproduce the exact value of one particular point. As already discussed, such a comparison does not consider concentration distributions occurring during simulations with accumulations of compositions close to c_0.

For a second comparison of the two methods, the absolute values of the differences between the original Gibbs energy and its approximations are calculated around different c_0 for the Taylor expansion $|\Delta G_T(c)| = |^{id}G(c) - G_T(c)|$ and also for the fit with the least squares method $|\Delta G_F(c)| = |^{id}G(c) - G_F(c)|$. Hereby, the ranges $c_1 \leq c \leq c_2$, from which twenty data points are taken for the fits, vary in size due to the choice of different ε to calculate $c_1 = c_0 - \varepsilon$ and $c_2 = c_0 + \varepsilon$. In figure 12.5(a) the deviations related to a Taylor approximation and a quadratic fit are plotted around $c_0 = 0.1$ with $\varepsilon = 0.05$. Not surprisingly the deviations of the Taylor expansion to the original function are smaller in the vicinity of c_0 than the deviations $|\Delta G_F(c)|$. For concentrations, which are further away from c_0, the functions fitted with the least squares method better resemble the original Gibbs energies. It can also be seen, that the errors due to the polynomial regression between c_1 and c_2 are distributed around the average value $\langle|\Delta G_F(c)|\rangle$, which is represented by a horizontal line. If the size of the fitting range is smaller due to a choice of $\varepsilon = 0.025$ like in figure 12.5(b), than also the range around c_0 is narrower, in which the deviations $|\Delta G_T(c)|$ are smaller than $|\Delta G_F(c)|$. While the parabola from the Taylor approximation perfectly resembles the original graph at c_0, the curve of $|\Delta G_F(c)|$ has three different roots in all of the plotted examples, which are located between c_1 and c_2. Due to this, there is a narrow region in between the inner region around c_0, for which $|\Delta G_F(c)| < |\Delta G_T(c)|$ holds. The errors are also calculated for $c_0 = 0.4$ and the resulting deviations are plotted in figures 12.5(c) and 12.5(d). As already discussed before, the differences for approximations close to $c = 0.5$ are smaller compared to the approximations close to the pure components. Despite this difference, the courses of the curves are similar to the ones for $c_0 = 0.1$. These results emphasize the fact, that Taylor approximations are better suited to reproduce the original functions at particular concentrations, such as the equilibrium compositions. If more widespread concentrations are expected, than the least squares method is an appropriate alternative to determine the polynomial coefficients.

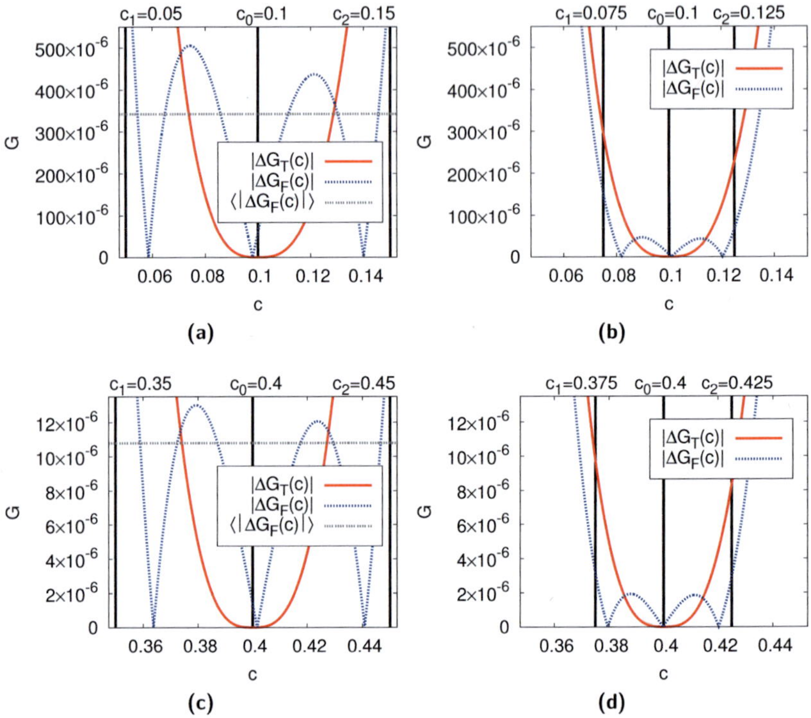

Figure 12.5: Comparison of the deviations for a Taylor expansion $|\Delta G_T(c)|$ and a least squares fit $|\Delta G_F(c)|$ in dependence of the composition c_0 and the fitting range $c_1 \leq c \leq c_2$. Close to c_0, around which the Taylor expansion is done, the differences $|\Delta G_T(c)|$ are smaller, while further away the functions from the least squares method better resemble the original Gibbs energies. In (a) the data for the fitting is taken from a broader range around $c_0 = 0.1$ defined by $\varepsilon = 0.05$. The value of $\langle|\Delta G_F(c)|\rangle$ from figure 12.4 is marked as a horizontal line. In (b) the data range is smaller with $\varepsilon = 0.025$. Subfigures (c) and (d) show the comparison of $|\Delta G_T(c)|$ and $|\Delta G_F(c)|$ for a choice of $c_0 = 0.4$. The fitting of $G_F(c)$ is performed in (c) for a data range of $0.35 \leq c \leq 0.45$ and in (d) for a range of $0.375 \leq c \leq 0.425$.

13 Conclusions and Outlook

In the previous sections, a framework for the coupling of thermodynamic datasets with the quantitative grand potential model is derived. The methodology relies on the fact that phase transformations in most technically relevant alloys occur close to equilibrium and only limited information is sufficient to describe the essential physics of the problem in the phase-field setup. In light of this fact, a reduction of the computational effort can be achieved by the use of simplified formulations instead of the thermodynamic functions from the database.

As a visualization of the derived preprocessing framework, the flowchart in figure 13.1 comprises the possible approaches to utilize thermodynamic and kinetic data for phase-field simulations. The source of all thermodynamic data are experiments and first-principle methods. In an assessment process, this data is collected, weighted and stored in thermodynamic datasets via Gibbs energies. By applying commercial thermodynamic software or self-developed code, the Gibbs energies can be tabulated for the relevant temperatures and compositions. Applying a suited formulation to approximately describe the functions, which are required for the phase-field solver, the Gibbs energy values can then be used to derive the unknown parameters (e.g. by Taylor expansion). To validate the results, the approximated equilibrium compositions and corresponding phase-diagrams can be recalculated by the thermodynamic software. In order to do this, the derived functions have to be read into the thermodynamic software via the standard TDB file format. Unfortunately, the syntax of this file format imposes restrictions on the applicable formulations, because coefficients of Redlich-Kister type are expected. This limits the usability of commercial software to recalculate phase diagrams from approximated descriptions and gives rise to workarounds. The application of self-developed code or open-source software like OpenCalphad [18] enables the incorporation of arbitrary formulations and should be considered as an alternative to commercial software. Within the presented framework, mobility data can also be used to derive diffusion coefficients, which can be treated as constants under the assumption of only small changes in temperature and composition. Additionally, the kinetic coefficients for the phase-fields can be calculated from the thermodynamic functions and the diffusion data according to equation 7.20.

As a consistent further development of the presented framework, one could design datasets specifically for the simulations they should be used for. They would only include the measured data of the temperature- and composition regimes, that are relevant for the phase transformations to be simulated. This would enable the usage of simple expressions, because the datasets do not have to cover the phase diagram in its entirety. By doing so, any refitting procedure would be obsolete. Ideally one could increase the amount of available data in the region of interest by additional targeted measurements. Figure 13.1 also includes this concept of assessments being optimized for simulative purposes.

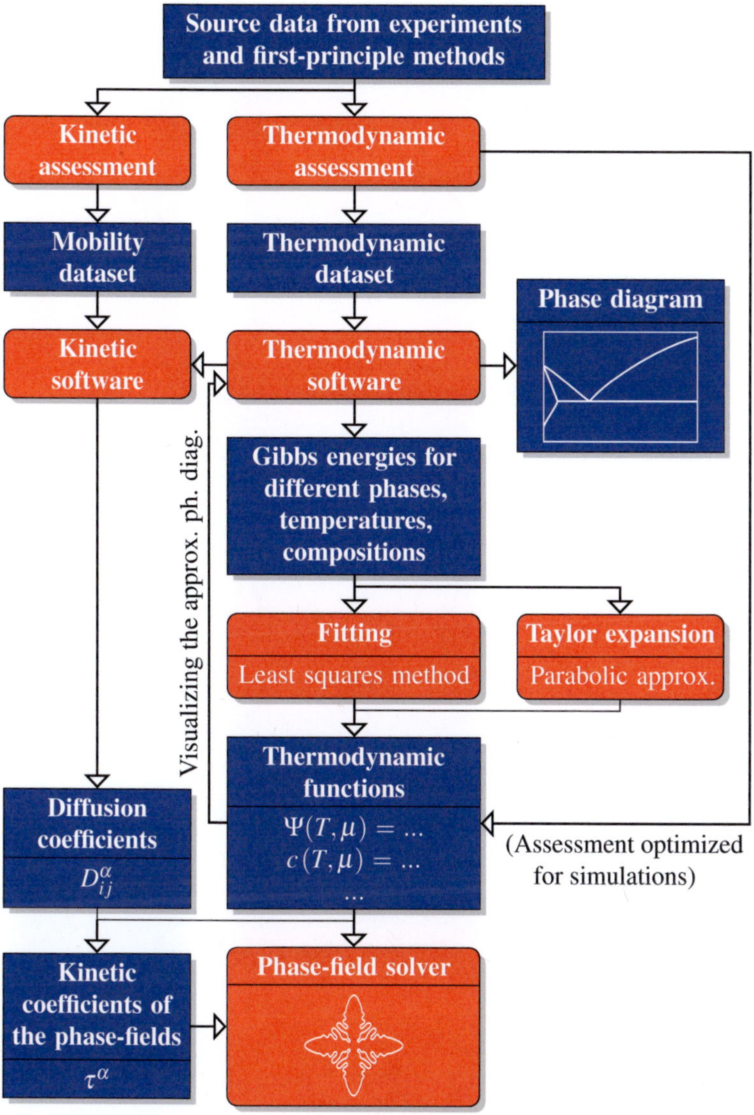

Figure 13.1: Flowchart of the derived preprocessing framework. Blue rectangles denote data, functions and diagrams. Red rounded rectangles represent software and actions.

Part IV

Phase-field studies based on CALPHAD data

The previous part comprises different approaches for the coupling of CALPHAD data with the phase-field model based on grand potentials. In the present part, phase-field studies are presented, showing that plausible results can be achieved through the utilization of thermodynamic datasets. The main topic of the studies is planar and dendritic solidification of aluminum-silicon in dependence of varying material and process parameters. These simulation results are compared to analytical models for solidification including a discussion about the assumptions underlying the models. Besides the solidification of Al-Si, this section also addresses solid state transformations in the system Fe-Cu.

14 Stability analysis of single-phase growth in Al-Si-(Mg) alloys

The eutectic system aluminum-silicon is one of the most important cast alloys based on aluminum [3, 4, 5]. A common third alloying element is magnesium, which improves the mechanical properties. However, the properties in the solid state are strongly dependent on the microstructural features, which are in turn influenced by the evolution processes during solidification. A prediction of in-situ microstructural evolution is therefore of great interest as it is a precursor to the prediction of the properties of the alloy. Modeling and simulation of solidification microstructures based on phase-field methods is hence very useful in determining the microstructural response to the given processing conditions. In doing so, it must be considered that casting is not a homogeneous process, as there are zones of purely equiaxed dendrites, while in other regions solidification happens in a columnar manner. To enhance the knowledge about solidification, its variation upon change in processing conditions and compositions have to be taken into account. A lot of investigation has been done for binary systems, however, the number of phase-field studies decreases with every additional component, understandably because of the increasing complexity and the absence of thermodynamic and mobility databases, which are requisite inputs in any sort of thermodynamic modeling.

Equiaxed dendritic growth is one of the first structures simulated with the phase-field method [41, 82, 83] and is intensively studied ever since. For Al-Si in particular, the equiaxed solidification of multiple dendrites under isothermal conditions is discussed by Zhang et al. [84]. Zhao and Hou [85] report simulations on equiaxed solidification of a single Al-Si dendrite, taking into account the influence of changing temperature. For Al-Si and Al-Cu, Ohsasa et al. [86] determine a correlation between the fractal dimension of equiaxed dendrites and the content of solute in phase-field studies, as well as in experiments. The influence of composition is considered by Zhang et al. [87] for ternary aluminum-based alloys. Their simulations of dendritic growth are carried out with a phase-field solver linked to thermodynamic databases. They observe that the addition of a ternary component can change the solidification velocity due to the influence on the diffusion behavior.

While equiaxed solidification proceeds equally in all directions, the crystals grow aligned in the case of columnar solidification. These morphologies evolve when small protrusions of planar fronts get amplified and start to form cellular arrangements. Such a transition from planar growth to shallow cells is analyzed by Bi and Sekerka [88]. A rich variety of patterns can emerge after the breakdown of the planar front. The morphological change from shallow to deep cells and the concomitant adjustment of the wavelengths is investigated by Lan, Shih and Lee [89]. Stable subunits of two cells, so called doublons, occur in simulations computed by Losert et al. [90]. Boettinger and Warren [54] observe a variety of structures in directional solidification, ranging from chaotic cells at low pulling speeds to regular cells and the reoccurrence of a planar front at higher velocities. In their phase-field study of directional solidification, Wang et al. [91] examine the selection mechanisms of primary dendritic spacing. After the planar instability, the microstructure passes a stage of seaweed-like growth and finally develops to an array of cells and columnar dendrites. A detailed work on dendritic to fractal structures for varying growth conditions is presented by Amoorezaei et al. [92] at the example of an Al-Mg alloy. For large spacings side arms can develop and the growth takes place as columnar dendrites. In a study on the Al-Cu system, Steinbach [93] describes, that the interface anisotropy strongly affects their spacing. Columnar dendrites are simulated for directionally solidifying Al-Si by Diepers and coworkers [94]. Their investigations suggest that different steady states can evolve depending upon the initial setting. Such a history dependence is also found by Amoorezaei et al. [95]. This emphasizes the relevance of understanding the onset of columnar growth, as it predetermines the conditions of later growth stages.

The Mullins-Sekerka theory [8] (see section 6.1) provides a criterion about the stability of perturbed planar fronts. This theory includes the destabilizing effects resulting from the concentration gradients in front of the interface together with the stabilizing effects due to surface energy and is therefore suited for a comparison with simulation results. Echebarria et al. [49] report on a good agreement of their phase-field simulations of directional solidification with the stability spectrum from MS. Badillo and Beckermann [96] perform a phase-field study of the columnar-to-equiaxed transition and validate their model by comparison with the MS predictions. A similar line of discussion is drawn in the present study for the planar-to-columnar transition. The MS theory relies on the assumption, that the system consists of only two components and has isotropic surface energies. The strength of anisotropy is however found to be an important factor influencing the stability of cellular arrangements, as discussed for 2D simulations by Kopczyński, Rappel and Karma [97] or for 3D simulations by Dejmek et al. [98] and by Ma et al. [99]. Following the derivations of Mullins and Sekerka, Coates et al. [100] derived a similar stability analysis for dilute ternary systems undergoing unidirectional solidification. A more general theory for an arbitrary number of components is provided by Hunziker [101]. Both theories assume a moving temperature gradient, which is used as a condition to derive the unknown parameters.

14.1 Thermodynamic functions for the binary simulations

The first part of the following solidification study deals with the pure binary system aluminum-silicon. To set up the simulations, the required thermodynamic functions are fitted with the least squares method (see also section 11.2) from CALPHAD data documented by Feufel et al [29]. Figure 14.1 shows the corresponding Al-Si phase diagram with the range of concentrations, for which the phase-field simulations are performed using the grand potential model. This range only covers a small part of the diagram, which illustrates the reason of modeling the thermodynamic properties by simplified expressions only for the relevant part of the system. The fits are carried out for a temperature of 875 K and the considered aluminum concentrations reach from $c = 0.9$ to $c = 0.995$. In the following, aluminum acts as the independent component and therefore c denotes the concentration of Al and the chemical potential μ refers to the partial derivative of the Gibbs energies with respect to the aluminum concentration. Being stable in the chosen concentration range, only the liquid phase and the aluminum rich FCC phase are regarded within the scope of this survey. In the assessment of Feufel et al., both phases are modeled according to the Redlich-Kister-Muggianu formulation [25, 26] with a single lattice, for which the end members are the pure elements. Hence the Gibbs energies are described with equations 5.5 to 5.7, whereby the Gibbs energies of formation and the interaction parameters of the FCC and liquid phase are listed in table 14.1.

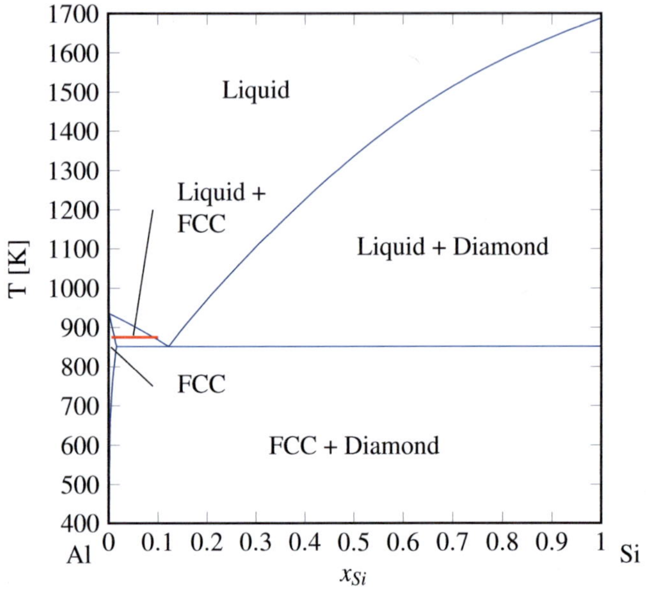

Figure 14.1: Only the hypo-eutectic part of the Al-Si phase-diagram at a single temperature is studied by phase-field simulations in this thesis and is represented by a red line.

As explained in section 12.1, it is beneficial from a computational point of view to use polynomials for the modeling of the thermodynamic functions. By applying quadratic polynomials for the free energy densities, the formulas of the grand potential densities and phase concentrations can be derived analytically as described in section 10.1. On the other hand, quadratic polynomials do not reproduce the original functions as good as polynomials of higher order. To fit the functions for the system Al-Si, an accuracy-focussed approach is used. Hereby the functions for the phase concentrations and grand potentials are fitted individually as polynomials of the chemical potentials and the polynomial degree is adjusted to derive an acceptably small error of the approximation.

At first, the functions of the phase concentrations for the FCC and the liquid phase are derived. As already described in section 11.2, the invertibility of the chemical potential functions has to be given in order to derive expressions for c^{α} as functions of μ. For the examined temperature and concentration range, this prerequisite is fulfilled, because the chemical potentials of both phases are strictly monotonic with concentration. A proof of the monotonicity can be found in appendix C. The functions for the phase concentrations are determined according to the procedure described in section 11.2 by fitting polynomials of degree 4 for data points of c over μ at the respective temperature for both of the phases. To quantify the error of the fits, the absolute values of the differences between each of the data points and the corresponding values calculated with the approximated functions are calculated for both phases. The average values of these deviations calculated over all data points can be found in table 14.2. With a maximal average deviation of less than 0.005%, a good match can be stated for the chosen approximations. The resulting fits at $T = 875$ K for both phases are plotted in figures 14.2(a) and 14.2(b), which also display the monotonicity of the functions.

Table 14.1: Expressions for the parameters used in the interpolation polynomials

Parameter	Temperature interval [K]	Formula (for temperatures in K, Gibbs energies in J/mol result)
$^{\circ}G_{Al}^{FCC}(T)$	700.0 - 933.6	$-11276.24 + 223.02695T + 74092T^{-1} + 18.531982 \cdot 10^{-3}T^2 - 5.764227 \cdot 10^{-6}T^3 - 38.5844296T\ln(T)$
$^{\circ}G_{Si}^{Dia.}(T)$	298.15 - 1687	$-8162.609 + 137.227259T + 176667T^{-1} - 1.912904 \cdot 10^{-3}T^2 - 0.003552 \cdot 10^{-6}T^3 - 22.8317533T\ln(T)$
$^{\circ}G_{Si}^{FCC}(T)$	298.15 - 6000	$51000.00 - 21.8T + {}^{\circ}G_{Si}^{Dia.}(T)$
$^{0}L_{Al,Si}^{FCC}(T)$	298.15 - 6000	$-3143.78 + 0.39297T$
$^{\circ}G_{Al}^{liq.}(T)$	298.15 - 933.6	$11005.553 - 11.840873T + 7.9401 \cdot 10^{-20}T^7 + {}^{\circ}G_{Al}^{FCC}(T)$
$^{\circ}G_{Si}^{liq.}(T)$	298.15 - 1687	$50696.4 - 30.0994T + 2.09307 \cdot 10^{-21}T^7 + {}^{\circ}G_{Si}^{Dia.}(T)$
$^{0}L_{Al,Si}^{liq.}(T)$	298.15 - 6000	$-11340.10 - 1.23394T$
$^{1}L_{Al,Si}^{liq.}(T)$	298.15 - 6000	$-3530.93 + 1.35993T$
$^{2}L_{Al,Si}^{liq.}(T)$	298.15 - 6000	2265.39

Table 14.2: Average deviations of the approximations

Phase α	$\Psi^\alpha(\mu)$	$c^\alpha(\mu)$
FCC	0.000865%	0.000769%
Liquid	0.000134%	0.004279%

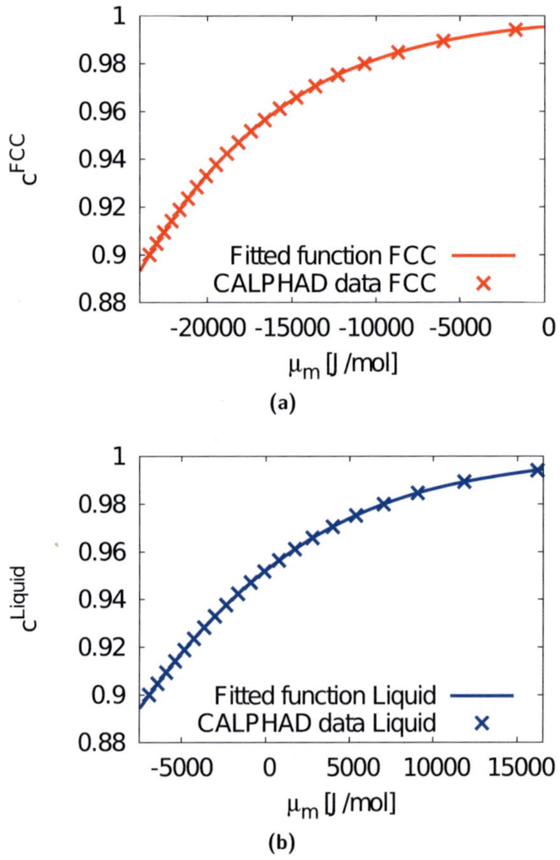

Figure 14.2: Based on CALPHAD data, functions for the phase concentrations are fitted over the molar chemical potential for (a) the FCC phase and (b) the liquid phase. To get functions dependent on density values, the molar chemical potentials have to be divided by the molar volume V_m.

By analytical differentiation, the functions for the partial derivatives of the phase concentrations with respect to the chemical potentials are determined from the polynomial formulations of $c^{\alpha}(\mu)$. To derive functions for the grand potentials, again ordered pairs of Ψ^{α} and μ from the chosen composition range are calculated for each phase. From these sets of data points, functions $\Psi^{\alpha}(\mu)$ are derived by least-squares calculations and again quartic polynomials match well with the data from CALPHAD. In figures 14.3 the results are plotted over the aluminum concentration.

In the used dataset, the FCC and liquid phase are described with the Redlich-Kister-Muggianu formulation, which does not include sublattices or additional physical contribution parts. As higher order polynomials are used to approximate the original data in an accurate manner, there is no marked reduction of complexity associated with the fitting of these phases. A refit of the CALPHAD assessment becomes more beneficial if, for example, the phases are described with sublattices, for which there is no explicit rule for the conversion of concentrations to site-fractions.

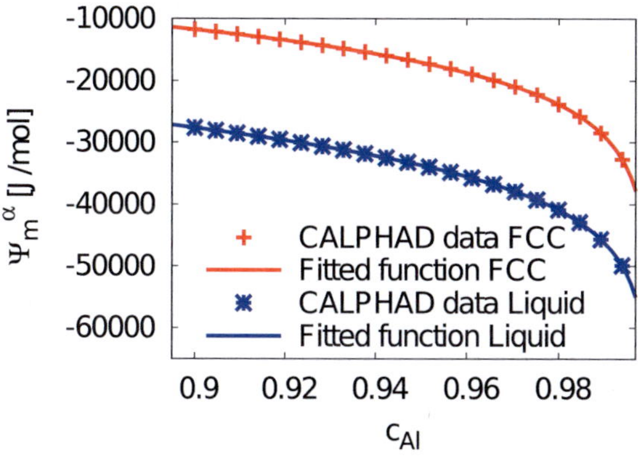

Figure 14.3: Molar grand potential fitted over chemical potential and plotted over concentration for FCC and liquid phase. To get functions for density values, the molar grand potentials have to be divided by the molar volume V_m.

14.2 Mullins Sekerka study of perturbed growth fronts

Utilizing the previously derived thermodynamic functions, the aluminum alloy is investigated in terms of the stability of planar growth fronts in the following. Starting from the amplification of small perturbations, such fronts can evolve to columnar dendrites or other growth morphologies. The stability analysis is performed with regards to different influencing factors, firstly the effect of anisotropic surface energies and secondly the influence of a third component in combination with varying diffusivities. As these conditions are not considered in the work of Mullins and Sekerka described in section 6.1, modifications of the classical theory are introduced in the following chapters.

14.2.1 Simulation study for isotropic and anisotropic surface energies

Simulation setup

While accurate data for the Gibbs energies is provided by the CALPHAD method, data for the surface energies and diffusivities are not available to the same extent. For the inter-diffusivities in the liquid a value of $D^{liq} = 3 \times 10^{-9}\ \mathrm{m}^2/\mathrm{s}$ reported for Al-Si in Kurz and Fisher [34] is taken. The diffusion coefficients in the solid phase are orders of magnitudes lower and an exact value is not decisive on the studied timescales. For that reason, the diffusivity in the FCC phase is set to be $1 \times 10^{-13}[m^2/s]$. To counterbalance the artificial solute trapping effect arising for the strongly different diffusivities on both sides of the diffuse solid-liquid interface, the simulations are carried out with an anti-trapping current (see also equation 7.22). At first isotropic surface energies of $\sigma = 0.225\ \mathrm{J}/\mathrm{m}^2$ are set, as it is assumed in the MS theory. The chosen value is calculated from data for the Gibbs-Thomson coefficient and entropy of fusion per unit volume from [34]. Then, the influence of the cubic crystal structure of the solid phase is studied by applying an anisotropy for the surface energies of the type given in equation 7.17. In [102], the strength of anisotropy is reported with the value $\delta_{\alpha\beta} = 0.01$ for Al-Cu. For Al-Si, a stronger anisotropy can be found in [103], which is measured as an aggregate value for all solid phases, including also the faceting diamond-cubic phase. As a distinct value for the Al-rich FCC phase cannot be found, the strength $\delta_{\alpha\beta}$ is varied by 0.01, 0.02 and 0.04. For the simulations presented in this paragraph, the interpolation function $h_\alpha^{III}(\phi)$ defined in equation 7.5 is chosen. Furthermore the relaxation constant τ is calculated in order to achieve vanishing interface kinetics (see also section 7.2.1).

The MS theory deals with the stability of a flat solidification front. The growth of such a planar front, can be reproduced appropriately by simulating solidification in a one dimensional domain. The domain length is set up to 750 computational cells with a cell size of $\Delta x = 2$ nm. The simulations are executed with a moving frame algorithm to efficiently simulate the microstructure evolution in a small window shifted in the direction of growth. The moving window algorithm is combined with the extrapolative boundary condition described in section 7.2.3. This condition is applied at the upper boundary and sets the chemical potentials in an extrapolative way to reproduce the diffusion profile in the liquid outside of the domain. Throughout the whole domain, a constant temperature of $T = 875$ K is prescribed. The thermodynamic functions are fitted as described in section 14.1. At the bottom of the domain, the FCC phase is seeded with its equilibrium chemical potentials at the considered temperature of 875 K. The rest of the domain is filled up with the liquid phase. In the binary system Al-Si, the chemical potential of the liquid phase corresponds to an aluminum concentration of 0.985. As the equilibrium concentration of the liquid is 0.912 for this temperature, the setup induces a supersaturation of the liquid phase, causing the solid to grow. For all simulations, a time step width of $\Delta t = 0.025$ ns is applied. The chosen parametrization is summarized in table 14.3.

Table 14.3: Parameters of the binary simulations.

Parameter	Symbol	Value
Domain height		750 cells
Cell size	Δx	2 nm
Time step width	Δt	0.025 ns
Temperature	T	875 K
Diffusivity of liquid	D^l	3×10^{-9} m^2/s [34]
Diffusivity of FCC	D^{FCC}	$1 \times 10^{-13}\ m^2/s$
Surface energy	σ	0.225 J/m^2 [34]
Initial liquid conc.	c^l_{Si}	0.015

When the front velocity between two time steps only changes by less than 1×10^{-5} % of the last velocity, the planar front is considered to grow at steady state and the simulation is stopped. To study the stability of the planar front, 2D simulations of sinusoidal solid-liquid interfaces are performed, as assumed in the MS theory. Such a growth front can be modeled in a simplified way by only calculating one or half a cellular crystal, as already discussed in e.g. [104, 96, 105, 91]. For a choice of the domain widths of $\lambda/2 = \pi/\omega$ and with mirror boundary conditions (zero Neumann boundary conditions) on the left and right side, this setup acts as a representative segment of a periodic front with a wavelength λ and a frequency ω. To study the stability of the planar front, the domain is initialized with the steady state profile and small sinusoidal ripples are applied, as assumed in the MS theory. To realize a sinusoidal shape, the domain is shifted into the growth direction by an offset $\Delta z = \delta \sin(\omega x)$, with x as the coordinate normal to the growth direction and δ as a small amplitude. This procedure ensures that the steady state diffusion profile obtained from a prior 1D simulation is shifted together with the solid-liquid interface, which is an assumption of the MS theory.

Stability analysis for isotropic and anisotropic surface energies

At first, isotropic surface energies are applied and the perturbed fronts are restarted for different wavelengths in the regime of the critical wavelength, at which the rate of change is zero. In figure 14.4 the amplitudes are plotted over time. For smaller wavelengths, the amplitude of the perturbation decreases and the front returns to the planar growth mode. Amplification of the ripple appears for larger wavelengths, indicating the transition from planar to cellular growth. The fastest increase in amplitude of the plotted wavelengths happens for $\lambda = 0.32$ µm.

The temporal change of amplitude at a time step n can be calculated by the forward difference $\dot{\delta}_n = \dfrac{\delta_{n+1} - \delta_n}{t_{n+1} - t_n}$. The normalized rates of change of the amplitudes $\dot{\delta}_n/\delta_n$ are averaged in the regime, where the amplitudes are changing linearly with time. These averaged rates for both, isotropic and anisotropic surface energies of different strengths are plotted in figure 14.5 together with the predictions from MS theory according to equation 6.7. As it can be seen, the analytical curve from MS theory is well reproduced. By interpolating between the measuring points, the critical frequency ω_0 and wavelength λ_0 of the phase-field simulations are determined. The results of both quantities ω_0 and λ_0 are listed in table 14.4 together with the predicted values of the MS theory.

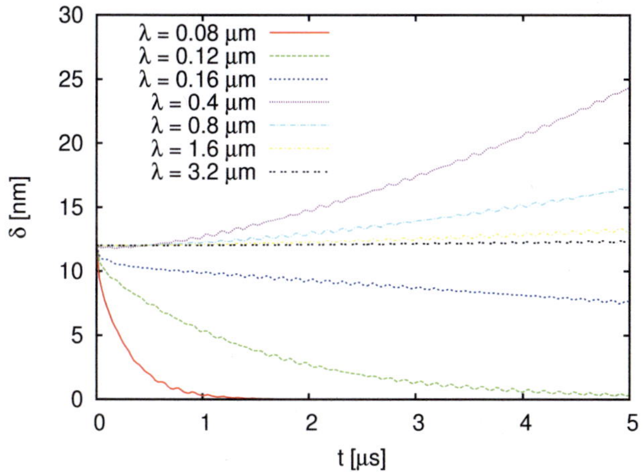

Figure 14.4: Absolute amplitudes plotted over time for isotropic surface energies.

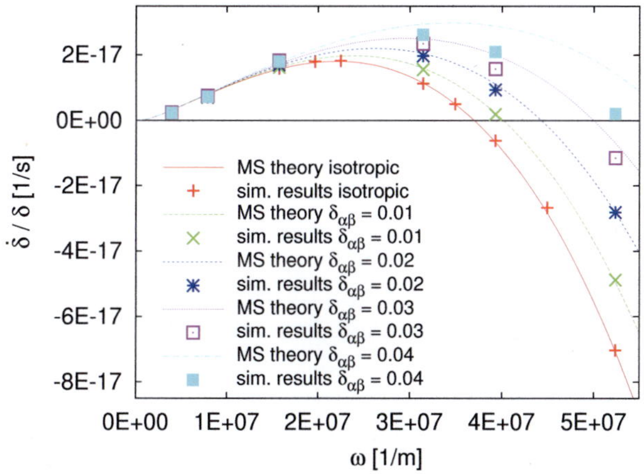

Figure 14.5: The rate of change of amplitude over the frequency for different strengths of anisotropy of the surface energy, ranging from an isotropic property to a strength of $\delta_{\alpha\beta} = 0.04$.

Table 14.4: Critical frequencies and wavelengths obtained from simulations and calculated from MS theory.

	ω_0 sim. $[\mathrm{m}^{-1}]$	ω_0 MS $[\mathrm{m}^{-1}]$	λ_0 sim. $[\mu\mathrm{m}]$	λ_0 MS $[\mu\mathrm{m}]$
isotropic	3.704×10^7	3.695×10^7	0.17	0.17
$\delta_{\alpha\beta} = 0.01$	3.998×10^7	4.01×10^7	0.157	0.157
$\delta_{\alpha\beta} = 0.02$	4.334×10^7	4.422×10^7	0.145	0.142
$\delta_{\alpha\beta} = 0.03$	4.815×10^7	4.991×10^7	0.131	0.126
$\delta_{\alpha\beta} = 0.04$	5.374×10^7	5.861×10^7	0.117	0.107

As a next aspect of the MS analysis, the effect of the surface energy anisotropy between the solid and liquid phase on the microstructure formation is investigated. Cubic anisotropies of different strengths $\delta_{\alpha\beta} = 0.01$, 0.02, 0.03 and 0.04 are applied to the surface energy with the same value of σ as in the isotropic case. The preferred growth direction due to anisotropy is chosen such that it coincides with the growth direction of the overall front. Figure 14.5 shows the results for the different strengths of anisotropy in comparison to the isotropic results. For the same initial wavelengths, higher rates of change are measured, when the strength of anisotropy is increased. Compared to isotropic interfaces, the stabilization of planar fronts due to the Gibbs-Thomson effect is reduced for anisotropic surface energies, thus decreasing the possible stable wavelengths.

There is no generalized MS theory for anisotropic growth. Nevertheless, anisotropy can be respected by replacing the surface energy σ by the effective stiffness of the solid-liquid interface $\sigma_{eff} = \sigma\left(1 - 15\delta_{\alpha\beta}\right)$, which accounts for one of the influences. In the range of bigger wavelengths, the measured rates of change for a small perturbation are nearly identical for isotropic and different anisotropic conditions and the predictions from MS theory. This is plausible as the influence of curvature undercooling is smaller for bigger wavelengths and thus the effect of anisotropy gets negligible. As can be seen in figure 14.5 and table 14.4, a good agreement between the theory and the simulation results is also found in the regime of smaller wavelengths for low strengths of anisotropy. However, with an increase of anisotropy, there is significant deviation between the simulation results and the analytical predictions for smaller wavelengths. An explanation is the difference in the interfacial shape in the simulations from that assumed in the linear-stability analysis performed by Mullins and Sekerka, thereby bringing in non-linear effects which get highlighted for smaller wavelengths, where the influence of curvature is larger.

Morphological changes due to anisotropy

The resulting growth patterns for the variation of anisotropy and the domain widths are shown in figure 14.6. For small wavelengths of the perturbations, the front stabilizes and returns to the planar growth mode, while for bigger wavelengths the perturbations amplify and cells develop. The images include wavelengths close to the critical ones. In these cases, the rate of change is so small, that the front preserves the imposed rippled shape until the end of the simulation. Tip splitting occurs for the cells with isotropic surface energy, while parabolic shapes appear in the case of anisotropy. With increasing anisotropy, the tips change continuously from being blunt to a more needle-like shape.

The simplified setting is well suited to study the stability of planar growth and the onset of cellular growth. Nevertheless this setup of simulating half a cell can only reproduce periodic fronts consisting of uniform cells. To avoid the resulting restrictions, a next set of simulations is performed in a broader domain, such that the front can freely develop to its favored morphology. The side length of the quadratic domain is set to 2500 computational cells, corresponding to a physical length of 5 μm. At the left and at the right side of the domain, periodic boundary conditions are chosen. Mirror conditions are imposed at the bottom and extrapolative boundary conditions are applied at the top. At the beginning of the simulation, solid phase is filled at the bottom of the domain, having the contour of superimposed sine waves with small amplitudes. The wavelengths are chosen as $\lambda_1 = 0.0628$ μm, $\lambda_2 = 0.214$ μm and $\lambda_3 = 0.415$ μm, such that the shortest wavelength λ_1 is smaller than all λ_0 from table 14.4.

The first simulation is conducted with isotropic surface energy. At the beginning, the cells evolving from the initial perturbations overgrow each other and their number reduces. The growth front develops into a compact seaweed structure, as can be seen in figure 14.7. Because of the missing anisotropy, the FCC phase grows erratically with a constant creation of new branches due to tip splitting. The resulting mean spacing $\overline{\lambda}$ of the simulated seaweed structure is 0.122 μm, which is 0.7 times the critical wavelength λ_0 from the simulations of the stability analysis for the isotropic case. In the cell borders, liquid droplets with an increased concentration of silicon are caught. Without imposing a velocity by a moving temperature gradient, the self-selected front velocities are a degree of freedom to be measured. The velocity of the seaweed structure is 0.73 m/s and is hence smaller than the front velocities for the anisotropic surface energies. All front velocities and mean spacings for the variation of the anisotropy are documented in table 14.5.

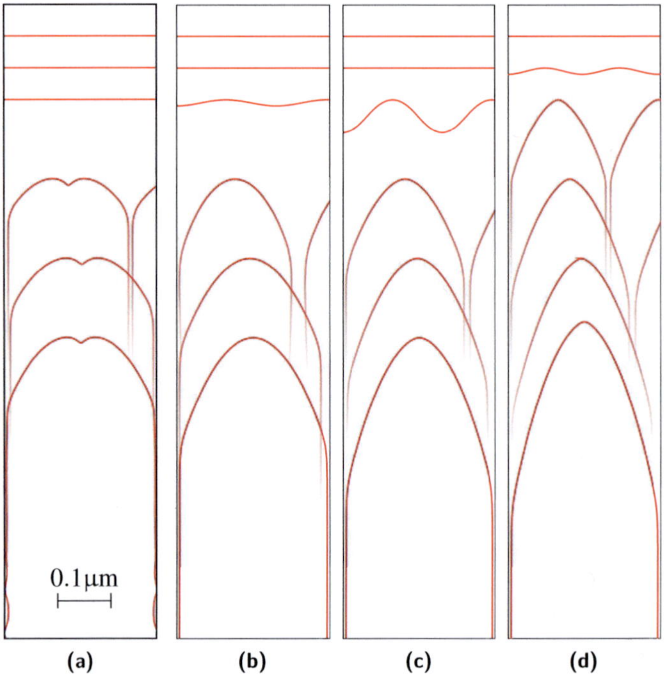

Figure 14.6: Contour lines of the solid-liquid interfaces for different strengths of the surface energy anisotropy and for different domain widths. Image (a) refers to isotropic surface energy, image (b) to an anisotropy of $\delta_{\alpha\beta} = 0.01$, (c) shows the contours for $\delta_{\alpha\beta} = 0.02$ and (d) for $\delta_{\alpha\beta} = 0.04$. In all images, the contours from bottom to top display the results observed for spacings of 0.28 µm, 0.24 µm, 0.2 µm, 0.16 µm, 0.12 µm and 0.08 µm. The pictured contours show the front at time $t = 25$ µs, when the initial perturbations have developed to the favored shape for these conditions. With increasing anisotropy, the cells change from more finger-like to more needle-like shapes.

For an anisotropy strength $\delta_{\alpha\beta} = 0.01$, the same setting as for the isotropic case leads to steadier growth (see figure 14.8), but still not as regular as for the strongest anisotropies. In an adjustment period at the beginning of the simulations, the initial number of cells reduces. After that, the surviving cells start to develop different growth morphologies. Some pairs of cells grow cooperatively in a doublon-like manner. In other regions of the front, the cells grow in oscillatory modes. Again, liquid inclusions occur between the cells. Similarly to the simulations with isotropic surface energies, the tips of the cells are not located on the same level into the growth direction. For example, some of the doublon-like pairs grow ahead of their neighboring tips.

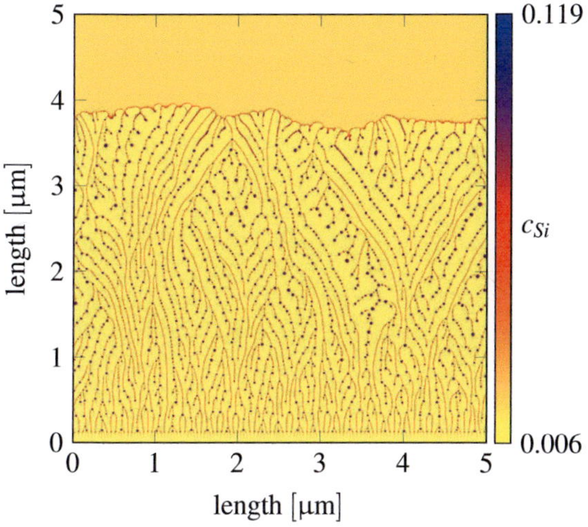

Figure 14.7: For isotropic surface energies, the initial perturbations develop to seaweed patterns with droplet inclusions in the solid structure.

For the simulations with anisotropies of $\delta_{\alpha\beta} = 0.02$ and $\delta_{\alpha\beta} = 0.04$, regular cellular growth occurs, as can be seen in figures 14.9 and 14.10. Once steady state is reached, the cells are oriented into the crystallographically favored direction with the tips being located at nearly the same level in this direction. Again, some of the liquid grooves separating neighboring cells transform into chains of droplets. No oscillations occur and the cell borders form straight lines. The mean spacings for the stronger anisotropies are smaller than the ones for $\delta_{\alpha\beta} = 0.01$

or isotropic conditions. While the cells for $\delta_{\alpha\beta} = 0.02$ have blunt tips and similar widths, the tips are more pointed for the strongest anisotropy and stronger deviations in the widths of the individual cells are observed.

Table 14.5: Velocities and mean spacings of the structure for a variation of the anisotropy. The mean spacings are compared to the critical spacings λ_0 derived from the simulations (see table 14.4).

anisotropy	front velocity	mean spacing $\bar{\lambda}$
isotropic	0.73 m/s	0.122 µm ($0.706 \times \lambda_0$)
$\delta_{\alpha\beta} = 0.01$	0.889 m/s	0.122 µm ($0.777 \times \lambda_0$)
$\delta_{\alpha\beta} = 0.02$	1.037 m/s	0.064 µm ($0.441 \times \lambda_0$)
$\delta_{\alpha\beta} = 0.04$	1.252 m/s	0.074 µm ($0.632 \times \lambda_0$)

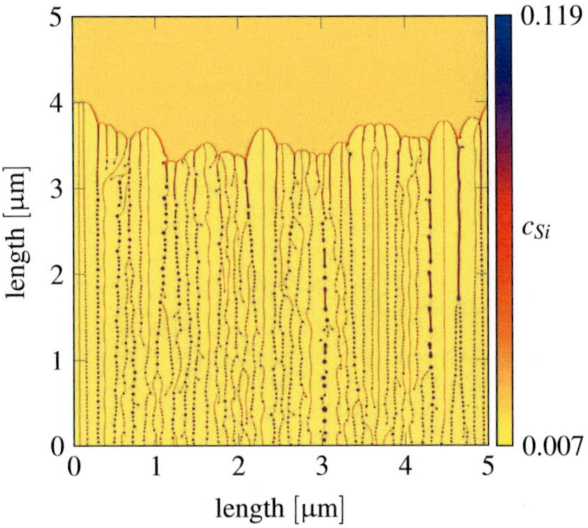

Figure 14.8: Non-uniform growth of cells for a weak anisotropy of the surface energy ($\delta_{\alpha\beta} = 0.01$).

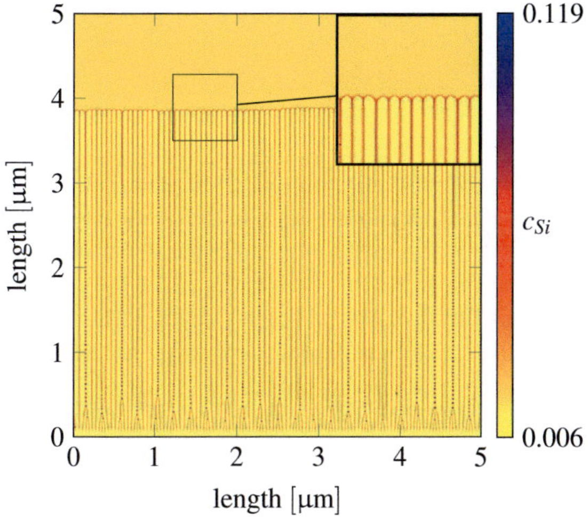

Figure 14.9: After an initial period of consolidation, the application of an anisotropy strength $\delta_{\alpha\beta} = 0.02$ leads to steady cellular growth. The cells of similar width grow parallel into the crystallographically favored direction.

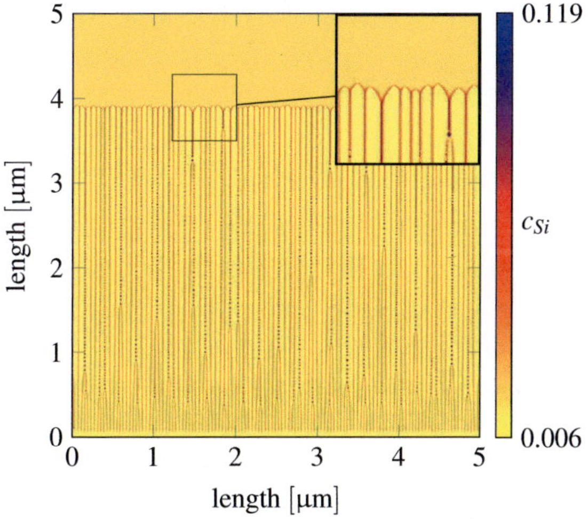

Figure 14.10: A surface energy anisotropy of $\delta_{\alpha\beta} = 0.04$ causes the perturbations to develop to regular cells. Similar as for the case of $\delta_{\alpha\beta} = 0.02$, the grooves turn into chains of droplets.

14.2.2 Simulation study of ternary impurity

As most of the relevant alloy systems consist of more than two components, the stability of perturbed fronts is studied also for ternary systems in the following. The classical MS theory is derived for binary material systems and thus not suited for a comparison with ternary simulation results. The stability analysis of Coates et al. [100] for dilute ternaries and also the analysis for multi-component systems from Hunziker [101] assume a moving temperature gradient to solve for the unknowns. For the isothermal conditions of the present simulation setup, this assumption is not fulfilled. Therefore, a modified MS criterion for ternary material systems with constant temperature is derived in the following, which is comparable to isothermal simulation results.

The classical MS theory provides an equation for the growth rates of the sinusoidal perturbations, which simplifies to equation 6.7 for isothermal conditions without a temperature gradient. In a ternary system, the same equation must be fulfilled for both solutes i:

$$\frac{\dot{\delta}}{\delta} = b_i \frac{D^l_{ii}}{c_i^0(1-k_i)} \left[k_{\omega,i} - \frac{v}{D^l_{ii}}(1-k_i) \right] + v \left(k_{\omega,i} - \frac{v}{D^l_{ii}} \right) \tag{14.1}$$

with the decay rate for a frequency ω given as

$$k_{\omega,i} = \frac{v}{2D^l_{ii}} + \sqrt{\left(\frac{v}{2D^l_{ii}}\right)^2 + \omega^2}. \tag{14.2}$$

While the diagonal coefficients of the diffusion matrix D^l_{ii} and the partition coefficients k_i are material parameters, the interface compositions of the planar front c_i^0 and the corresponding steady state velocity v are process parameters, which can easily be derived from the simulation results. The parameters b_i, are used to approximate the composition fields c_i^ϕ at a perturbed interface by $c_i^\phi = c_i^0 + b_i \delta \sin(\omega x)$ and need to be solved for in the following manner. As equation 14.1 has to return the same rates of change for both of the two solutes, a relation between b_1 and b_2 can be found by subtracting equation 14.1 with the parameters of the first solute from equation 14.1 with the parameters of the second solute. The ternary Gibbs-Thomson equation provides a second relation for the coefficients

$$m_1 b_1 + m_2 b_2 = \Gamma \omega^2 \tag{14.3}$$

with m_i as the liquidus slopes. In the present analysis, the ternary Gibbs-Thomson coefficient for the solid liquid interface is interpolated as

$$\Gamma = \frac{\sigma \sqrt{\sum_{i=1}^{2} m_i^2}}{\sqrt{\sum_{j=1}^{2} \left(\sum_{i=1}^{2} \frac{\partial^2 f^s}{\partial c_i \partial c_j} \Delta c_i \right)^2}} \tag{14.4}$$

using the difference of the equilibrium concentrations of the solid and the liquid Δc_i and the second derivative of the solid free energy with respect to the concentration. Given these relations, all unknowns are fixed and the extended theory can be applied for the following stability analysis.

Simulation setup

To analyze the influence of a ternary component on the growth rates and compare the modified stability theory with phase-field simulations, small amounts of the alloying element magnesium are added. For the liquid phase, the same silicon concentration of $c_{Si} = 0.015$ as in the binary case is used and the compositions of Mg and Al are varied. Like for the binary simulations, the thermodynamic data for both phases at the Al rich corner of the system Al-Mg-Si is provided by the assessment of Feufel et al. [29]. The ternary dataset is utilized at isothermal conditions with a description of the free energy densities as concentration dependent parabolas, according to equation 10.6. For the temperature of 875 K, the functions $f^\alpha(\mathbf{c})$ are constructed around the equilibrium concentrations given by certain tie-lines between the solid and liquid phase. The coefficients are determined via Taylor expansion, by matching the first and second derivatives of the quadratic free energies with the ones of the Gibbs energies from the database and using its information about the grand potentials. This method for the determination of the thermodynamic coefficients is described in section 11.1. In a first simulation series, the liquid concentration of Mg is increased by 0.005 towards the liquid composition in the binary case, while the aluminum is decreased by the same amount. The tie-line for the thermodynamic approximation of this supersaturated liquid composition is shown in figure 14.11. It is chosen such that it goes through a point close to $c_{Al} = 0.98$, $c_{Mg} = 0.005$, $c_{Si} = 0.015$. A second simulation series is set up for the same amounts of magnesium and silicon $c_{Mg} = c_{Si} = 0.015$. The corresponding tie-line is also shown in figure 14.11. For both tie-lines, the solidus and liquidus lines are recalculated from the fitted polynomials and visualized in the ternary diagram. As one would expect, the equilibrium compositions are well reproduced in the vicinity of the chosen tie-lines. Like in the binary simulations, the chemical potential fields in the solid phase are initialized with the equilibrium values for the respective tie-line. The chemical potentials in the liquid phase correspond to the supersaturated composition.

To begin with, the interdiffusivities in the solid and liquid phase are chosen to be comparable to the binary setup. This means, that in the case of liquid the value of $3 \times 10^{-9}[m^2/s]$ is applied for the diagonal entries of the ternary diffusion matrix. More accurate diffusivities are calculated using the same approach as Zhang et al. [106]. For deriving the mobilities, one can take the impurity diffusion coefficients of Mg and Si in liquid aluminum together with the self-diffusion coefficient of liquid aluminum (see table 14.6). As described in [32], these impurity- and self-diffusivities can be used to derive the inter-diffusivities D^l_{ij} of the two solutes in the liquid phase. This calculation requires thermodynamic factors, which can be calculated from the thermodynamic database. For the composition $c_{Al} = 0.97965$, $c_{Mg} = 0.00535$, $c_{Si} = 0.015$, the diffusivity matrix in table 14.7 is derived. For the sake of comparability, the same diffusivities are applied for both compositions under consideration. For all simulations, the diffusivity matrix of the solid phase only has diagonal entries with the value of $1 \times 10^{-13}[m^2/s]$. The surface energy is chosen to be isotropic, while all other material parameters and growth conditions are identical to the binary runs (see table 14.3).

Stability analysis for a third component and different diffusivities

In the first series, a value of $3 \times 10^{-9}[m^2/s]$ is applied on both diagonal entries of the diffusion matrix while the off-diagonal entries are zero. For each composition, a one-dimensional simulation is carried out. At the time, when the growth velocity reaches the steady-state value from the binary case, the simulation is stopped. The same perturbations as previously are applied on the ternary fronts by using the setting described in section 14.2.1 and from the 2D simulations, the rates of change are derived. Figure 14.12 shows the simulation results of the ternary systems with equal diagonal diffusivities, which are compared to the binary outcomes and the modified stability theory. With an increasing amount of magnesium, the critical frequencies and the maxima of the curves decrease. For all compositions, a good match can be found between the analytical curves and the simulation results.

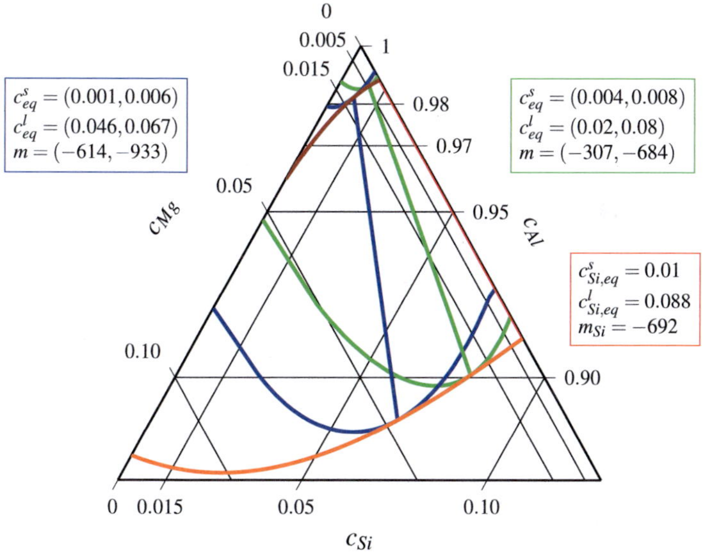

Figure 14.11: The Al-rich corner of the ternary phase-diagram at 875 K together with the material parameters needed in the MS anlysis. The liquidus slopes are given in $[K/\text{mole fraction}]$ and the vectors denote (c_{Mg},c_{Si}). The liquidus line is shown in orange and the solidus line in brown. A tie-line (green color) is chosen, such that it goes through a point close to $c_{Al} = 0.98, c_{Mg} = 0.005, c_{Si} = 0.015$. A second tie-line close to the point $c_{Al} = 0.97, c_{Mg} = 0.015, c_{Si} = 0.015$ is shown in blue color. The tie-line used for the binary simulations is shown in red. The figure also includes the solidus and liquidus lines calculated from the polynomial approximations, which are drawn in the same colors as the corresponding tie-lines.

Table 14.6: Self- and impurity-diffusivities of the liquid phase with R as the ideal gas constant and T as the temperature in K.

	Diffusivities $[m^2/s]$
D^*_{Al} from [107]	$1.16 \times 10^{-7} \exp(-21330/(RT))$
D^*_{Mg} from [108]	$9.9 \times 10^{-5} \exp(-71600/(RT))$
D^*_{Si} from [108]	$5.12 \times 10^{-8} \exp(-22200/(RT))$

Table 14.7: Inter-diffusivities D^l_{ij} of the liquid phase, calculated from the coefficients in table 14.6.

	Inter-diffusivities $[m^2/s]$
D^l_{MgMg}	5.245×10^{-9}
D^l_{MgSi}	-0.137×10^{-9}
D^l_{SiMg}	-0.191×10^{-9}
D^l_{SiSi}	2.568×10^{-9}

As a further stability study, unequal coefficients are applied on the diagonal of the diffusivity matrix for the liquid phase. These entries are varied according to the second column of table 14.8. Using the same procedure as in the previous studies, simulations are performed for both compositions under consideration. The different entries on the diagonal of the diffusivity matrix lead to rates of change as shown in figure 14.13. For the diffusivities under consideration, higher values of D^l_{SiSi} result in smaller critical frequencies as documented in table 14.8. The higher amount of magnesium leads to a narrower range of critical frequencies and also decreases the maxima of the curves. For both compositions, the analytical predictions match well with the simulation results.

For a last comparison, fully occupied diffusivity matrices are used in the simulation setups. Because the previously derived stability theory does not account for interactions due to non-zero off-diagonal elements in the diffusion matrix, these setups should result in deviations between the simulations and the analytical predictions. At first, the simulations are performed with the previously calculated diffusivity matrix documented in table 14.7 for the two compositions under consideration. As can be seen in figure 14.14, the rates of change from the simulations are close to the theoretical curves for the corresponding diagonal matrix. This result is plausible, as the off-diagonal diffusivity values are one order of magnitude smaller than the diagonal entries and the effect of diffusional interaction is thus small. For a composition of $c_{Mg} = c_{Si} = 0.015$, the curve for the fully occupied diffusion matrix is close to the curve for the diagonal matrix with equal coefficients.

To investigate the effect of diffusional interaction in greater depth, the diagonal entries of the diffusivity matrix are set to 3×10^{-9} m^2/s and positive values as well as negative values are applied for the off-diagonal entries. The matrices together with the critical frequencies and wavelengths can be found in table 14.9 and the results are displayed in figure 14.14. Analogously to the previous simulations, a higher amount of magnesium decreases the maxima of the curves and results in lower critical frequencies. The chosen positive cross-coupling coefficients correspond to repulsive diffusional interaction [101], which increases the effective diffusivity compared to the diagonal matrices. The increase of the effective diffusivity results in a bigger diffusion length, which in turn lowers the critical frequency. For the composition $c_{Mg} = 0.005$, $c_{Si} = 0.015$, the resulting rates of change and the critical frequency are slightly smaller than the analytical predictions for solely diagonal entries.

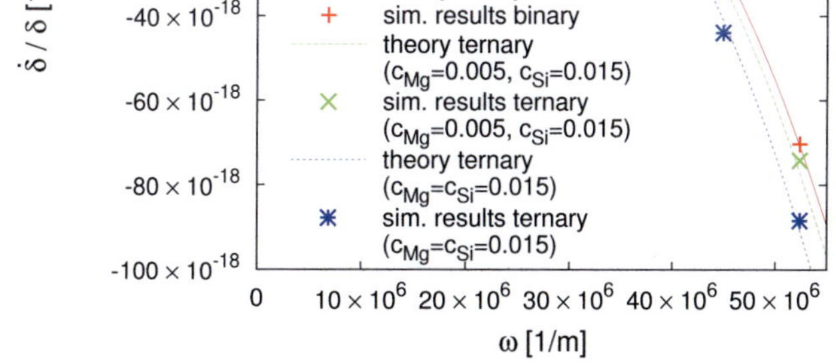

Figure 14.12: Growth rates from simulations and the analytical solutions for different amounts of magnesium with a constant concentration of silicon $c_{Si} = 0.015$ and equal diagonal diffusivities.

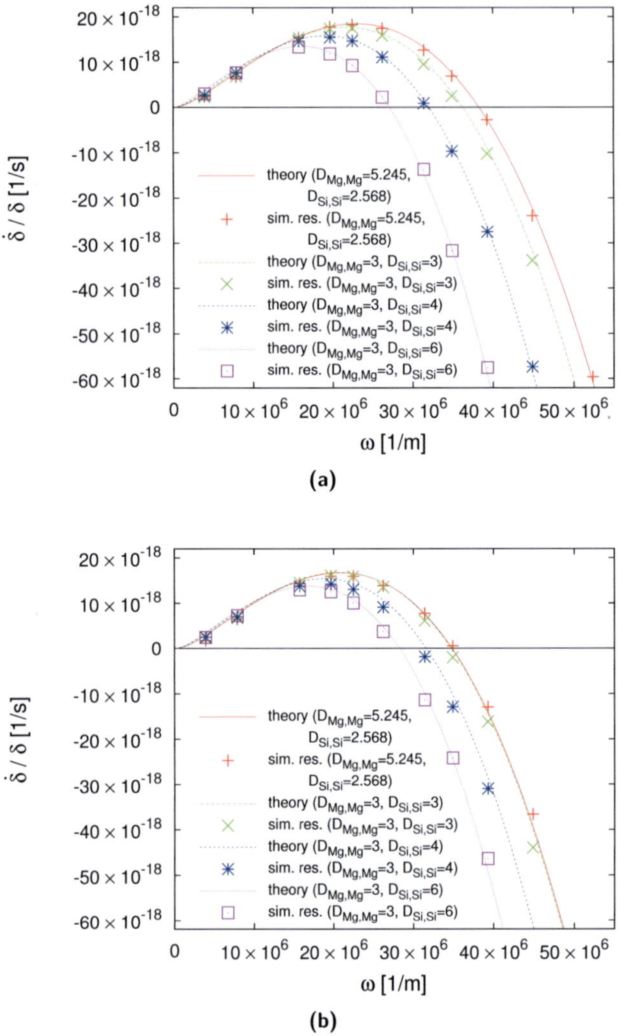

Figure 14.13: The rates of change for different entries on the diagonal of the diffusivity matrix of the liquid phase for the compositions (a) $c_{Mg} = 0.005$, $c_{Si} = 0.015$ and (b) $c_{Mg} = c_{Si} = 0.015$. (The diffusion coefficients in the diagram are nondimensionalized with a factor of 10^{-9} m^2/s.)

Table 14.8: Critical frequencies and wavelengths obtained from simulations and calculated from MS theory. The entries of the matrices denote (D^l_{MgMg}, D^l_{MgSi}, D^l_{SiMg}, D^l_{SiSi}) (nondimensionalized with 10^{-9} [m^2/s]). The values of the fourth and last line are taken from the calculated inter-diffusivities in table 14.7.

c_{Mg}	diffusion matrix	ω_0 sim. [m^{-1}]	ω_0 MS [m^{-1}]	λ_0 sim. [μm]	λ_0 MS [μm]
0.005	(3,0,0,3)	3.587×10^7	3.609×10^7	0.175	0.174
0.005	(3,0,0,4)	3.177×10^7	3.186×10^7	0.197	0.197
0.005	(3,0,0,6)	2.738×10^7	2.696×10^7	0.229	0.233
0.005	(5.245,0, 0,2.568)	3.818×10^7	3.825×10^7	0.164	0.164
0.015	(3,0,0,3)	3.413×10^7	3.45×10^7	0.184	0.182
0.015	(3,0,0,4)	3.099×10^7	3.153×10^7	0.203	0.199
0.015	(3,0,0,6)	2.775×10^7	2.81×10^7	0.226	0.224
0.015	(5.245,0, 0,2.568)	3.503×10^7	3.471×10^7	0.179	0.181

Table 14.9: Critical frequencies and wavelengths obtained from simulations with fully occupied diffusion matrices. The entries of the matrices denote (D^l_{MgMg}, D^l_{MgSi}, D^l_{SiMg}, D^l_{SiSi}) (nondimensionalized with 10^{-9} [m^2/s]).

c_{Mg}	diffusion matrix	ω_0 sim. [m^{-1}]	λ_0 sim. [μm]
0.005	(3,1,1,3)	3.599×10^7	0.175
0.005	(3,-1,-1,3)	4.252×10^7	0.148
0.005	(5.245,-0.137,-0.191,2.568)	3.847×10^7	0.163
0.015	(3,1,1,3)	1.113×10^7	0.564
0.015	(3,-1,-1,3)	3.807×10^7	0.165
0.015	(5.245,-0.137,-0.191,2.568)	3.597×10^7	0.174

A choice of negative cross-coupling coefficients, corresponding to attractive interactions [101], leads to considerably higher rates of change and a bigger critical frequency. For a composition of $c_{Mg} = c_{Si} = 0.015$, again the rates of change from simulations with positive off-diagonal coefficients are smaller than the theoretical values for the case without diffusional interaction. The simulated rates for negative off-diagonal coefficients are bigger than the predictions for the diagonal matrix in the range of larger frequencies. At smaller frequencies the rates of change for attractive interaction are smaller than the ones without interaction.

Morphological changes due to a third component

In the same manner as for the simulation series with varied anisotropy, the ternary simulations are carried out in larger domains again. The system size, the boundary conditions and the shape of the solid phase at the start of the simulation are identical with the settings for the binary system. Each simulation having a composition of $c_{Mg} = 0.005$ and $c_{Si} = 0.015$ starts with the growth of compact seaweed patterns. In a transient period, the seaweed structures turn into arrays of doublons, which are characterized by splitted tips. After growing in a curved manner, the doublons take a fixed orientation of 45° towards the growth direction of the overall front. The differences of the diffusion matrices do not lead to different growth morphologies. However, the velocity and the mean spacing of the doublon arrays is influenced, as documented in table 14.10. While the smallest growth velocities are measured for the diffusion matrices with equal diagonal entries and with positive cross-coupling coefficients, the front grows the fastest for the matrix with $D^l_{SiSi} = 6 \times 10^{-9}$ m^2/s. The biggest mean spacing is found for the matrix with $D^l_{SiSi} = 4 \times 10^{-9}$ m^2/s and the smallest mean spacing for the fully occupied matrix.

The growth as doublons is also preferred for a composition of $c_{Mg} = 0.015$ and $c_{Si} = 0.015$ for all of the diffusion matrices under consideration. At the beginning of each simulation, the biggest wavelength of the initial profile gets amplified, whereas the smallest wavelength vanishes. As a result, the evolving seaweed patterns are separated by broader regions of liquid and their distances are bigger compared to the simulations with less magnesium. In the final stage, the seaweed structures vanished and only tilted doublons remain, which start to develop side arms. As for all diffusivities only two doublons remain in the whole simulation domain, all of the arrays have the same mean spacing of 2.5 µm. Similar to the previous simulations, the biggest velocity is observed for the matrix with $D^l_{SiSi} = 6 \times 10^{-9}$ m^2/s, while all of the front velocities are less than half as big as the ones observed for the smaller magnesium content.

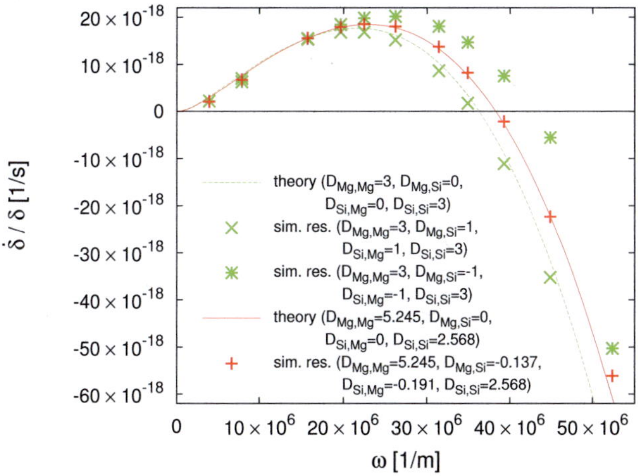

(a)

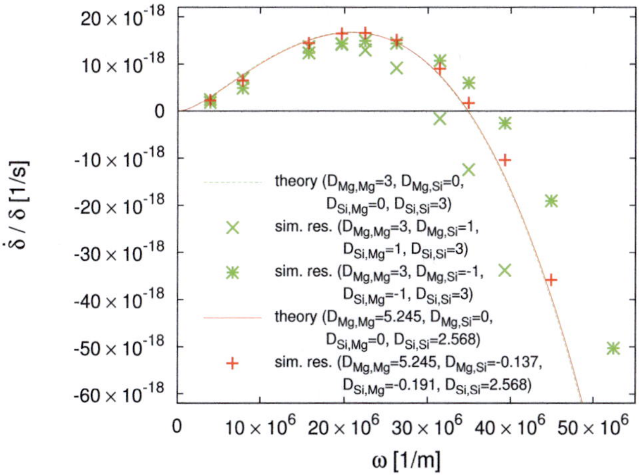

(b)

Figure 14.14: The rates of change for fully occupied diffusivity matrices and compositions of (a) $c_{Mg} = 0.005$, $c_{Si} = 0.015$ and (b) $c_{Mg} = c_{Si} = 0.015$. The results are compared with the theoretical predictions for the corresponding case of pure diagonal diffusivity. (The diffusion coefficients in the diagram are nondimensionalized with a factor of 10^{-9} m^2/s.)

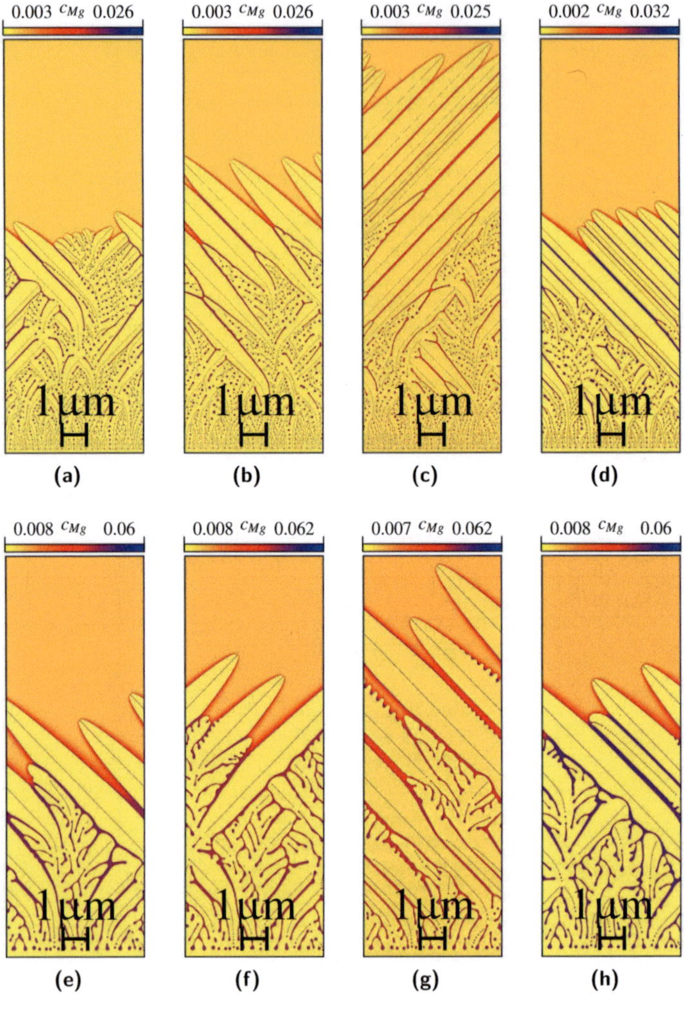

Figure 14.15: First line: Growth of doublons for the composition $c_{Mg} = 0.005$, $c_{Si} = 0.015$ (at 20 µs after the start) and diffusivity matrices of (a) (3,0,0,3), (b) (3,0,0,4), (c) (3,0,0,6), (d) (3,1,1,3). Second line: Growth of doublons for a magnesium content $c_{Mg} = c_{Si} = 0.015$ (at 50 µs after the start) and diffusivity matrices of (e) (3,0,0,3), (f) (3,0,0,4), (g) (3,0,0,6), (h) (3,1,1,3). The entries of the matrices denote (D^l_{MgMg}, D^l_{MgSi}, D^l_{SiMg}, D^l_{SiSi}) and are nondimensionalized with 10^{-9} [m^2/s].

Table 14.10: Mean spacings and velocities for a variation of composition and diffusivity. The entries of the matrices denote $(D^l_{MgMg}, D^l_{MgSi}, D^l_{SiMg}, D^l_{SiSi})$ (nondimensionalized with 10^{-9} [m^2/s]).

c_{Mg}	diffusion matrix	mean spacing $\overline{\lambda}$	velocity
0.005	(3,0,0,3)	1 μm	0.421 m/s
0.005	(3,0,0,4)	1.667 μm	0.5215 m/s
0.005	(3,0,0,6)	1 μm	0.7145 m/s
0.005	(3,1,1,3)	0.833 μm	0.438 m/s
0.015	(3,0,0,3)	2.5 μm	0.184 m/s
0.015	(3,0,0,4)	2.5 μm	0.211 m/s
0.015	(3,0,0,6)	2.5 μm	0.273 m/s
0.015	(3,1,1,3)	2.5 μm	0.203 m/s

14.2.3 Conclusions

In this section, the influence of both anisotropy and impurity composition on the stability of a planar interface is investigated for Al-Si alloys. In order to do this, the thermodynamic database for the particular alloy is effectively coupled to the grand-potential phase-field model. To account for the influence of anisotropic surface energies, the stability criterion from MS can be modified by using the effective stiffness of the solid-liquid interface. For small anisotropies, the dispersion relation from the modified MS theory matches well with the results from phase-field simulations. However, there is a concomitant change in the shape of the interface for stronger anisotropies, which introduces non-linear corrections to the curvature undercooling. The present modification of the linear stability analysis does not account for this influence. Hence, bigger deviations between the simulation results and the modified MS theory are observed for the higher strengths of anisotropy.

In the simplified 2D setup, the evolving cells form a blunt shape for a weak anisotropy, whereas needle-like structures are observed for a strong anisotropy. This is consistent with the outcomes for deep cells simulated in 3D by Gurevich et al. [109]. In an enlarged simulation domain, the morphological changes due to the anisotropy are visible. For isotropic surface energies, the front evolves to a compact seaweed pattern. Similar morphologies for low anisotropies have been obtained in phase-field simulations by Li and Yang [110]. A transition from seaweed structures in the isotropic case to straight cellular growth appears for increasing anisotropic strength. A similar transition due to low anisotropy is observed for three dimensional cellular growth by Ma and Plapp [99]. The occurence of seaweed patterns for low strengths of anisotropy is also found by Amoorezaei et al. [92]. In all of the binary simulations, liquid droplets are caught between neighboring cells. Such inclusions of solute in the deep regions of the grooves are observed in several 2D phase-field studies [89, 110, 105, 54] and in 3D by Ma and Plapp [99]. The resulting mean spacings $\overline{\lambda}$ for the enlarged domains are smaller than the corresponding critical wavelengths λ_0 from the simulations of the stability analysis.

Phase-field simulations are also performed to calculate the stability diagram upon addition of a ternary impurity, i.e magnesium. For the isothermal conditions used in the simulations, a modified stability criterion for ternary systems is derived. This theory takes account of different coefficients on the diagonal of the diffusion matrix. A good match is found for the comparison between the theory and simulation results for these kind of diffusion matrices. In the case of Al-Mg-Si, the off-diagonal entries of the diffusion matrix are one order of magnitude smaller than the diagonal ones. For such systems of relatively small diffusional interaction, the stability theory gives adequate predictions for the rates of change. If there are considerable cross-coupling effects, the growth rates are effected in a complex way, which cannot be predicted by the presented stability model. An extension of the current theory to account for diffusional cross-couplings and additional components is thus a topic for future work.

The influence of a third component on the evolving microstructure is studied in larger simulation setups. Starting from a compact seaweed structure in the binary case, the grooves between the branches widen with increasing concentration of Mg. Therefore, the mean spacings of all ternary simulations are larger than the corresponding critical spacings. This is in contrast to the binary simulations with varied anisotropies, which grow in a compact manner. The decrease of the compactness for higher amounts of Mg is connected to a decrease of the growth velocity, which is in accordance with the findings of Amoorezaei and coworkers [92]. In the current study, a variation of the diffusivities has only a small influence on the growth morphologies, as all of the ternary simulations result in tilted doublons.

In summary, it can be stated that the quantitative phase-field model and an extension of the classical Mullins-Sekerka theory result in the same stability characteristics of perturbed growth fronts. As the variation of the surface energy anisotropy, the composition and the diffusion coefficients can have a strong influence on the growth rates and morphologies, particular attention has to be paid for these factors in future studies.

14.3 Phase-field simulation of equiaxed dendritic growth

14.3.1 Simulation setup

In addition to the previous results of directional solidification, this section addresses the simulation of freely growing dendrites with four-fold surface energy anisotropy in a uniformly undercooled melt. In contrast to the previous study, only the binary system Al-Si is considered. The first series of simulations is performed in two-dimensional domains with 1000×1000 cells. Starting from a circular nucleus of the FCC phase at one corner of the domain and imposing mirror boundary conditions in the required coordinates, only one quarter of the dendrite is simulated. The cell size is chosen as $\Delta x = 0.05$ μm and the time step width as $\Delta t = 0.1$ μs, while the diffusion coefficients and surface energies are identical to the binary MS study in section 14.2.1. A cubic anisotropy of $\delta_{\alpha\beta} = 0.01$ is applied for the surface energy, according to the value reported for Al-Cu [102]. The whole domain is initialized with the equilibrium chemical potential for 900 K. In the same way as explained in section 14.1, the thermodynamic functions are fitted for a temperature of 894 K and the simulation is carried out for the latter temperature to derive a melt undercooling of $\Delta T = 6$ K.

14.3.2 Comparison to analytical predictions from LGK theory

Simulations of freely growing dendrites are performed and the measured velocities v and radii r are compared with those predicted from the analytical LGK theory [9], which is introduced in section 6.2. An input parameter required in the theory is the marginal stability criterion σ^*, which is calculated as $\frac{2d_0 D^{liq}}{vr^2}$ according to equation 6.11. As described in section 6.2, the tip radii are measured together with the velocities at steady state from the 2D simulation results in order to derive σ^*. The analytical tip radii calculated from σ^* are listed in table 14.11 together with the radii from the simulations. The maintained values from the simulations lie in between the two different analytical results given by LGK_{2D} (the parabolic plate solution proposed by Horvay and Cahn [36]) and LGK_{3D} (based on the Ivantsov integral [35] assuming a paraboloid of revolution). In the same way, the measured growth velocity differs from the LGK predictions, given in table 14.12. The differences of the 2D results to the outcomes of LGK_{3D} are expected due to the paraboloid approximation. To analyze the differences to the LGK_{2D} theory, different anisotropy strengths $\delta_{\alpha\beta} = 0.02, 0.04$ are considered similar as in the MS study. The variation is performed for the same undercooling of 6 K and the results are shown in figure 14.16. For the smallest anisotropy, the dendrite has a blunt tip and starts to form sidearms causing the deviation from the needle crystal shape assumed in the LGK_{2D} (plate) solution. With stronger anisotropy, the resulting radii tend towards the two-dimensional needle crystal solution. The radii are documented in table 14.11, the velocities can be found in table 14.12 and a visualization of the radii together with the contours of the evolving dendrites is given in figure 14.17.

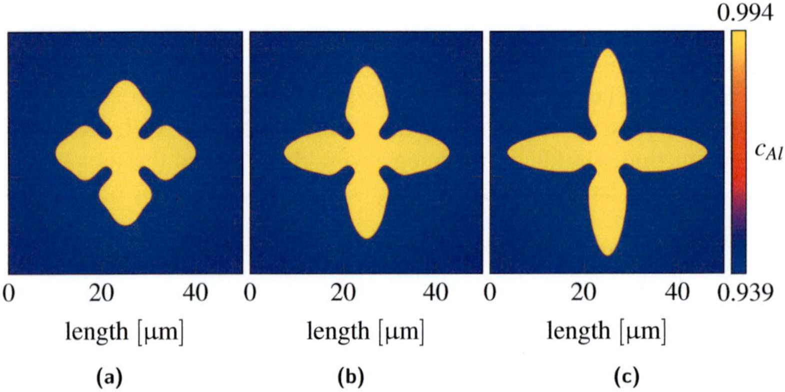

Figure 14.16: Equiaxed dendrites for a strength of anisotropy of (a) $\delta_{\alpha\beta} = 0.01$, (b) $\delta_{\alpha\beta} = 0.02$ and (c) $\delta_{\alpha\beta} = 0.04$. The plots show the concentration of aluminum.

As the LGK_{3D} theory considers the dendrite tip to be a three-dimensional object, simulations in 3D are conducted. The anisotropies are varied for the same alloy composition and undercooling as previously. For the different anisotropies, the domain sizes and time step widths are adjusted, according to table 14.13. Analogous to the previous simulations, a spherical nucleus is initiated at one edge of the domain, having mirror boundary conditions. The results together with the theoretical predictions are listed in tables 14.14 and 14.15 and are plotted in figure 14.18. As one would expect, the LGK_{3D} solution is in better agreement with the measured quantities than the 2D plate solution. To analyze the deviation between the assumed and the resulting shape of the dendrite, the solid-liquid-interfaces of cuts through the main arms 20 cells below the tips of the three dendrites are plotted in figure 14.19. It can be seen that the normalized tip radius for $\delta_{\alpha\beta} = 0.01$ is close to the unit circle, referring to a paraboloid of revolution. Due to the nature of the surface energy anisotropy, the dendrites cease to be pure paraboloids of revolution as assumed in LGK_{3D} theory. The cross-sectional shapes get more and more elliptical for the stronger anisotropy. However, this mismatch seems to be of minor importance as the best agreement between theory and results can be found for $\delta_{\alpha\beta} = 0.04$.

Table 14.11: Tip radii from LGK predictions and from 2D simulations.

$\delta_{\alpha\beta}$	σ^*	rad. for LGK_{2D}	rad. of 2D sim.	rad. for LGK_{3D}
0.01	0.095	4.667 µm	2.15 µm	0.717 µm
0.02	0.222	0.955 µm	1.14 µm	0.155 µm
0.04	0.349	0.617 µm	0.612 µm	0.105 µm

Table 14.12: Velocities from LGK predictions and from 2D simulations.

$\delta_{\alpha\beta}$	vel. for LGK_{2D}	vel. of 2D sim.	vel. for LGK_{3D}
0.01	0.0128 mm/s	0.0701 mm/s	0.544 mm/s
0.02	0.0616 mm/s	0.0875 mm/s	2.357 mm/s
0.04	0.0939 mm/s	0.111 mm/s	3.272 mm/s

In summary, LGK theory based on the approximation by Ivantsov gives good predictions of dendritic tip velocities and radii, for certain regimes in the parameter space of three-dimensional simulations. The LGK theory based on the plate approximation by Horvay and Cahn is suited for 2D simulations, which are close to the idealized shapes. A stronger surface energy anisotropy is resulting in dendritic shapes which resemble closely to the ones assumed in theory.

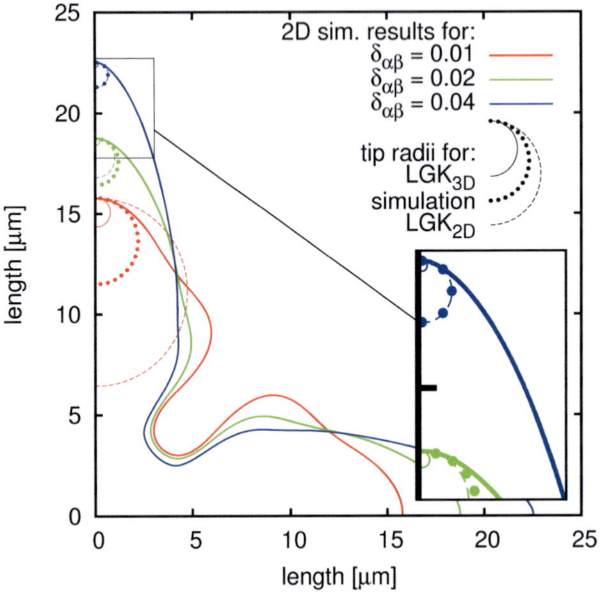

Figure 14.17: Contour plots of 2D simulation results together with radii from analytical LGK_{2D} and LGK_{3D} prediction. The dendrites are simulated with an undercooling of $\Delta T = 6$ K and different amplitudes of anisotropy. Due to the symmetry conditions only one quarter of each dendrite is simulated. The formation of a sidearm can be seen at the blunt tip of the dendrite with the lowest $\delta_{\alpha\beta}$. For higher $\delta_{\alpha\beta}$ the dendrite tips are more acute and no sidearm formation can be seen for the displayed growth stages.

Table 14.13: Simulation parameters for 3D.

$\delta_{\alpha\beta}$	domain side length	Δx	Δt
0.01	600 cells	0.04 μm	0.06 μs
0.02	600 cells	0.04 μm	0.075 μs
0.04	500 cells	0.04 μm	0.06 μs

Table 14.14: Tip radii from LGK predictions and from 3D simulations.

$\delta_{\alpha\beta}$	σ^*	rad. for LGK_{2D}	rad. of 3D sim.	rad. for LGK_{3D}
0.01	0.088	4.436 μm	0.685 μm	0.769 μm
0.02	0.149	2.467 μm	0.393 μm	0.389 μm
0.04	0.149	1.41 μm	0.227 μm	0.222 μm

Table 14.15: Growth velocities from LGK predictions and from 3D simulations.

$\delta_{\alpha\beta}$	vel. for LGK_{2D}	vel. of 3D sim.	vel. for LGK_{3D}
0.01	0.0134 mm/s	0.739 mm/s	0.509 mm/s
0.02	0.024 mm/s	1.101 mm/s	0.976 mm/s
0.04	0.042 mm/s	1.887 mm/s	1.708 mm/s

14.3.3 Conclusions

The present simulation study deals with the growth of binary equiaxed dendrites and the comparison of the resulting tip radii and velocities to the predictions from the established LGK theory. Agreement with the theory is found if the spatial dimensions of the simulation setup and the analytical solution are identical and the dendrites resemble ideal needle crystals. The simulations in 2D and 3D show a good match with the analytical values for high surface energy anisotropies. This condition contrasts to the MS theory, which is based on the assumption of isotropic surface energies. The mismatch between the anisotropic dendrites, having elliptical cross-sections in 3D, and the paraboloids of revolution assumed in LGK_{3D} theory is of minor importance. Hence, it can be stated that the LGK theories are better suited to describe the growth behavior of strongly anisotropic materials.

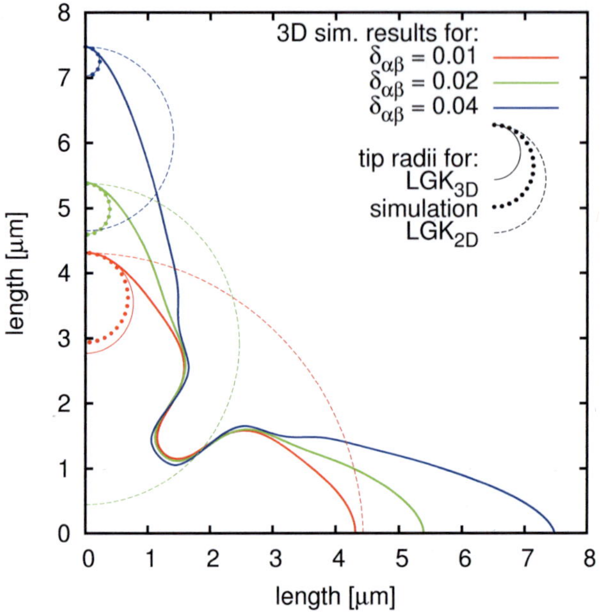

Figure 14.18: Sections through dendrites simulated in 3D with different $\delta_{\alpha\beta}$, whereby the sectional plane is congruent with one of the planes of symmetry. The measured tip radii are plotted together with the predicted radii from LGK_{2D} and LGK_{3D} theory.

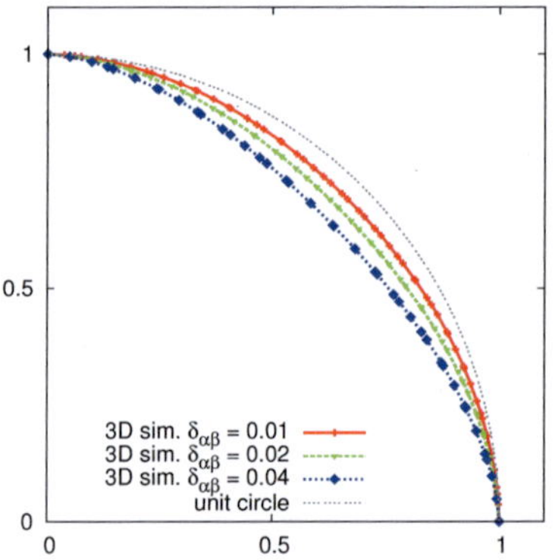

Figure 14.19: Cross sections through the main arms of the 3D dendrites compared to the unit circle. The contours of the different simulation results are scaled, such that their maximal radius is unity. With higher strengths of anisotropy, the deviation to a paraboloid of revolution (represented by the unit circle) is increasing.

15 Ostwald ripening of Fe-Cu – Influence of diffusion coefficients

In the following, an example for the application of diffusion data from computational kinetics (see section 5.2) is given, which also illustrates how to cope with Gibbs energy curves having two minima. The presented work is a continuation of a multiscale study on the coarsening behavior of Cu-rich precipitates in an α-iron matrix for isothermally aged Cu-Fe alloys [68] and a study on the coarsening of the ternary system Cu-Fe-Ni [111]. Within the framework of the multiscale study, the nucleation and early stages of precipitate growth are simulated by the kinetic Monte Carlo method (KMC) in 3D. After a step of conversion, the phase-field method is applied to continue with the later stages of growth on a bigger length scale. The resulting coarsening rates from a series of simulations performed for different compositions at a temperature of 1100 K are in good agreement with the analytical predictions from the Lifshitz-Slyozov-Wagner law (LSW) [112, 113]. In the subsequent investigation of Cu-Fe-Ni, the focus is laid on the composition paths of the matrix and the precipitates during coarsening in a 2D domain with an initial Gaussian distribution of circular precipitates. The objective of the present work is to account for the strong dependence of the coarsening process on the inter-diffusivities, which are decisive for its kinetics. In order to do this, the strength of diffusion is varied for identical starting configurations of the Cu-Fe alloy by applying different temperatures of 1000 K, 1050 K and 1100 K.

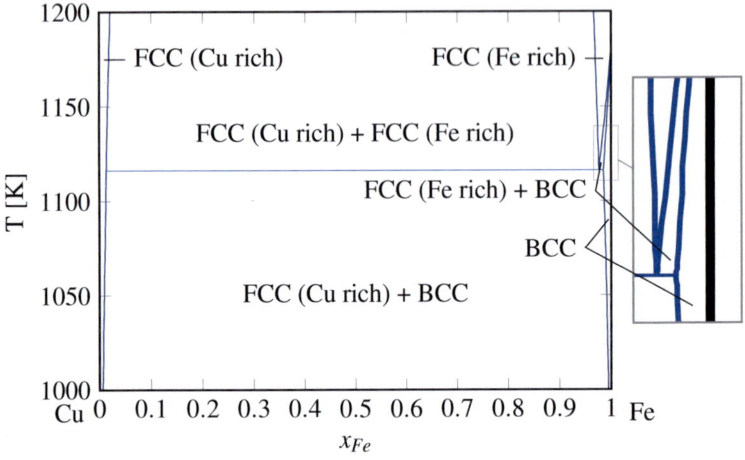

Figure 15.1: Phase-diagram of Cu-Fe in the temperature range from 1000 K to 1200 K.

15.1 Simulation setup based on data from computational thermodynamics and kinetics

Like in the previous studies, the thermodynamic data for the simulation setup is obtained from the CALPHAD method. The Gibbs energies are extracted from a thermodynamic assessment of Cu-Fe-Ni from Servant [114] (including magnetic contributions, as described in section 5.1.8). According to the phase-diagram shown in figure 15.1, a Cu-rich precipitate would have an FCC structure at the studied temperatures, but experiments have shown that the Cu-rich clusters first nucleate with a BCC structure and transform into FCC with increasing size [68]. This phenomenon can be explained with the nucleation advantage due to low interfacial energies for both phases having the same crystal structure. The molecular dynamics simulations carried out for the multiscale study justify this assumption, that for the considered particle sizes the matrix and precipitates are still perfectly coherent. Therefore, the Cu-rich precipitate phase, as well as the Fe-rich matrix phase, are identified in the current study with the BCC phase from the dataset. A plot of its Gibbs energies at the different temperatures is given in figure 15.2.

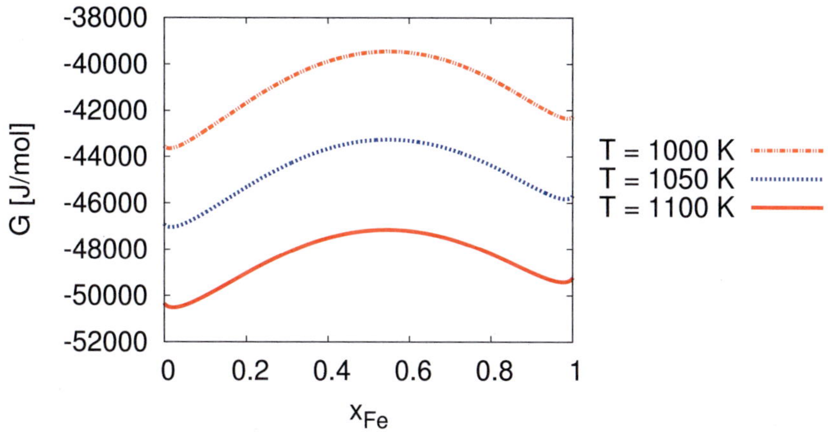

Figure 15.2: The Gibbs energies of the BCC phase from the CALPHAD dataset are plotted for the three different temperatures applied in this study. In every case the curves have a minimum at the Cu-rich side and a second one at the Fe-rich side.

As it can be seen, the Gibbs energy curves of the BCC phase exhibit two different minima for all plotted temperatures. If the Gibbs energy for a certain temperature would be described by a single expression in the phase-field setup, the resulting functions of the chemical potentials would not be invertible. As pointed out before, this invertibility is a prerequisite to determine the phase concentrations. Therefore, the matrix and the precipitates are described as separate phases in the present phase-field setup. Similarly to the treatment in the previous publications [68, 111], the free energies of the Fe-rich and the Cu-rich side of the BCC phase are fitted individually by isothermal quadratic formulations given by equation 10.1. In doing so, the coefficients for each temperature are determined by the least-squares method from data points around the equilibrium compositions. Such a fitting is sufficient because in the coarsening regime the compositions of the precipitate and matrix phase have already reached their equilibrium values with slight deviations due to curvature, which provides the driving force for coarsening. The resulting coefficients are documented in table 15.1 and the individual fits for $T = 1100$ K are graphically compared to the original free energy curves in figure 15.3. To convert the molar quantities from CALPHAD into densities, a value of $1 \times 10^{-6}\mathrm{m}^3/\mathrm{mol}$ is applied for the molar volume V_m.

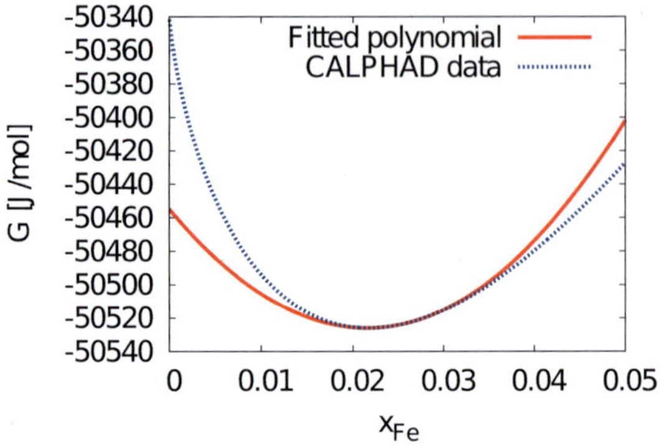

(a)

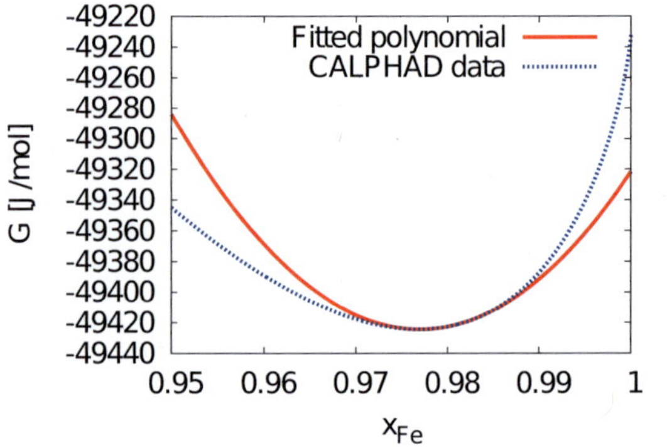

(b)

Figure 15.3: The Gibbs energy of the BCC phase in the Cu-Fe system from the CALPHAD dataset is plotted together with the approximation for $T = 1100$ K. The Gibbs energies of the Cu-rich (a) and Fe-rich side (b) are fitted independently as parabolas around the specific equilibrium concentrations.

In [68] it is derived, that the preferred interfaces of the Cu clusters in the Fe-rich matrix have an orientation of [110] and the corresponding surface energy is obtained from the broken bond model (BBM) [115, 116]. This surface energy of $\sigma = 0.41\ \mathrm{J/m^2}$ is also used in the present study. In the previous investigations, the inter-diffusivity is approximated from self diffusivities found in literature by making use of Darken's analysis. For the temperature of 1100 K the inter-diffusivity results as $1 \times 10^{-16}\ \mathrm{m^2/s}$. In the present study, the inter-diffusivities are derived from computational kinetics using mobility coefficients for iron and copper from the Landolt-Börnstein data collection (group III volume 26) [117] and the assessments of Jönsson [118, 119]. In combination with the thermodynamic dataset from Servant et al. [114], the diffusivities are calculated with the DICTRA module of the Thermo-Calc software for the three considered temperatures and are listed in table 15.1. In doing so, the composition of the Fe-rich matrix is used to calculate the inter-diffusivity of the BCC phase and this diffusion coefficient is also assigned to the precipitate phase. The relaxation constant τ is calculated according to equation 7.20 and can be found in table 15.1 for each temperature.

In contrast to the multiscale study, the initial setup of the simulation domains is not obtained from KMC results. Instead, the 2D domains are initialized like in the ternary study [111] with equal side lengths of 0.43 μm and filled with circular particles (see figure 15.4(a)). The Gaussian distribution of the particle radii of the present simulation series is the same as in the ternary study with the peak of the distribution at a radius of 2.87 nm and an average distance between the particles of 7.166 nm. Resulting from the CALPHAD database, the compositions for the Cu-rich BCC phase in equilibrium with the Fe-rich BCC phase are given in table 15.1. At the start of the simulation, the Cu-rich precipitates are filled with their equilibrium compositions, whereas the Fe-rich matrix is initialized with a small super-saturation of copper being documented in table 15.1.

Table 15.1: Material and phase-field parameters of the system Cu-Fe for different temperatures. (equi. means equilibrium, supersat. means supersaturated)

Temperature [K]	1000	1050	1100
Inter-diffusivity D $[m^2/s]$	4.289×10^{-17}	3.666×10^{-16}	1.953×10^{-15}
A^{BCC} (Fe-rich) $[J/mol]$	387018	255435	193341
B^{BCC} (Fe-rich) $[J/mol]$	-765091	-502150	-377768
X^{BCC} (Fe-rich) $[J/mol]$	335762	200962	135105
A^{BCC} (Cu-rich) $[J/mol]$	250946	189590	153324
B^{BCC} (Cu-rich) $[J/mol]$	-6252.69	-6229.37	-6604.28
X^{BCC} (Cu-rich) $[J/mol]$	-43620.6	-47002.5	-50454.9
c_{Fe}^{BCC} (Fe-rich), equi.	0.99	0.985	0.98
c_{Fe}^{BCC} (Fe-rich), supersat.	0.985	0.98	0.975
c_{Fe}^{BCC} (Cu-rich), equi.	0.015	0.02	0.025
τ $[Js/m^4]$	2.673×10^{21}	2.126×10^{20}	6.89×10^{18}

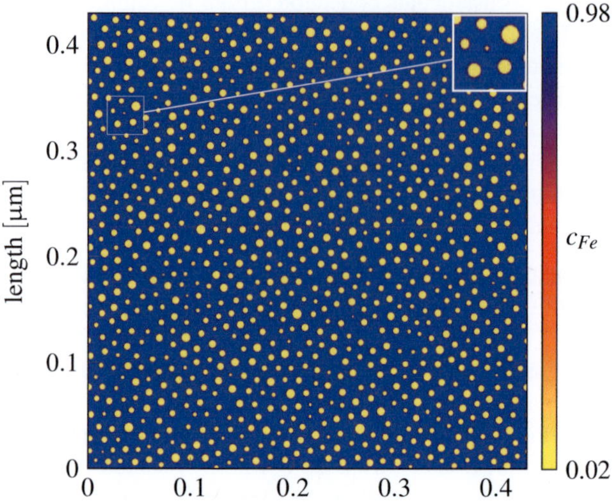

(a)

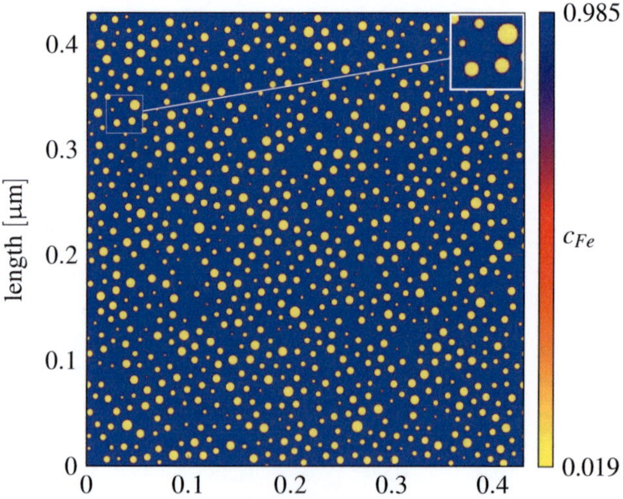

(b)

Figure 15.4: Microstructures for 1050 K in a 2D domain (0.43 µm × 0.43 µm). (a) The domain at the start of the simulation. (b) The microstructure after a coarsening period of 2660 s. While the smaller particles vanish, the initially bigger ones continue to grow.

15.2 Simulation series for different diffusivities

Due to the set supersaturation, a growth of all the precipitates takes place at the beginning of the simulation. After that initial stage, the process of Ostwald ripening begins and the larger precipitates grow in expense of smaller ones. In consequence, the number of particles decreases, which can be seen in figure 15.4(b) for the temperature of 1050 K. This coarsening process can be described by the established theory from Lifshitz, Slyozov [112] and Wagner [113]. The LSW theory predicts that independently from the initial configuration, different systems transition into statistically similar coarsening states. For this long-time regime, during which the coarsening takes place at steady-state, the theory predicts a linear increase of the average radius cubed r_{avg}^3 according to the formula

$$r_{avg}^3(t) = r_{avg}^3(0) + K_{LSW}t. \tag{15.1}$$

In this equation, $r_{avg}^3(0)$ denotes the average radius cubed at the beginning of the long-time regime and K_{LSW} is the coarsening rate constant, which depends on diffusivity, interfacial energy and solubility. In figure 15.5 it can be seen, that the coarsening for $T = 1100$ K is divided into three regimes. Immediately after the simulation start, the average radius increases, while the number of particles stays constant. This behavior is caused by the initial supersaturation and is connected to an increase of the phase fraction of the precipitates. After that initial period, the phase fraction of the precipitate phase stays constant and a stepwise increase of the average radius takes place due to the denucleation of particles. In this transient regime, the curve asymptotically approaches the steady-state course. The long-time regime is characterized by a linear increase of r_{avg}^3, which is in accordance with equation 15.1. For a measured rate constant $K_{LSW} = 0.023\ \text{nm}^3/\text{s}$ and $r_{avg}^3(0) = 22.127\ \text{nm}^3$ with the beginning of the long-time regime identified at $t = 225$ s, the linear K_{LSW} fit is also displayed in figure 15.5.

The different diffusivities, resulting from a variation of temperature, influence the time scales, at which the ripening takes place. In figure 15.6(a) the cube of the average radius is plotted over time for each simulation. For the higher temperatures and diffusivities, the r_{avg}^3 increases much faster. In a publication of Voorhees [120], a nondimensional time

$$\bar{t} = \left(Dc^s_\infty V_m / l_c^2\right) t \tag{15.2}$$

is proposed with c^s_∞ as the solute concentration in the matrix at a flat interface and the capillary length

$$l_c = 2\sigma V_m / (RT). \tag{15.3}$$

If c^s_∞ is identified with the supersaturated iron concentration and the radii cubed are plotted over the nondimensional time, the curves appear much closer, as displayed in figure 15.6(b). It has to be mentioned, that in difference to the paper of Voorhees, the concentration is given in mole fractions and therefore the molar volume does not have to be used in the calculation. As it can be seen, the curves for the lower diffusivities also asymptotically approach a linear growth regime.

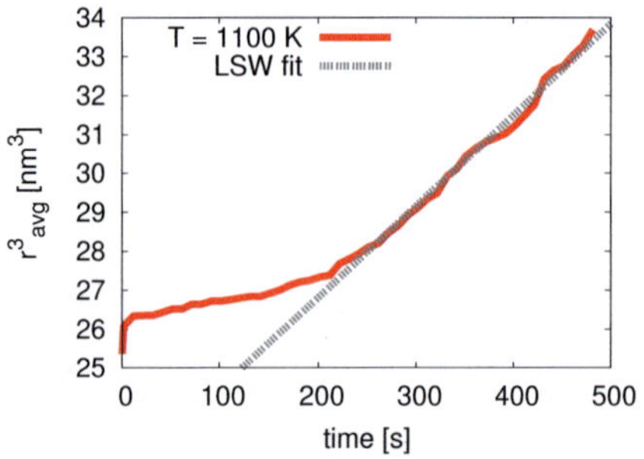

Figure 15.5: The cube of the average precipitate radii as a function of time for $T = 1100$ K. In the initial period ($t < 2$ s) the r^3_{avg} increases due to the supersaturation. In a transient period, the curve asymptotically approaches the steady-state growth, which starts at around 225 s. The coarsening kinetics according to LSW theory are drawn in grey.

Another aspect of the LSW theory is the analysis of the number of particles in different size classes in the long-time limit. In figure 15.7 the normalized numbers of precipitates are plotted over the radii scaled with the average radius. The discrete data for the temperatures of 1000 K and 1100 K at different simulation times is represented in 25 size classes. The simulations for different temperatures share the same initial Gaussian distribution. At steady-state the amount of precipitates with a radius close to the average radius is smaller than at the start of the simulation and in consequence the steady-state distributions are broader than the initial ones. In the figure it can be seen, that the distributions from different simulation times in the steady-state regime are self-similar. This so called dynamic scaling behavior is an important aspect of the LSW theory. Although dynamic scaling behavior is observed, the resulting distributions for all temperatures are broader than the curve predicted by the LSW theory. Distributions from experiments differ in the same way from the LSW predictions [121, 122], as well as the results from the previous phase-field simulations [111]. These differences are discussed in the paper about the multiscale approach [68] and can be explained by the assumptions made in the LSW theory.

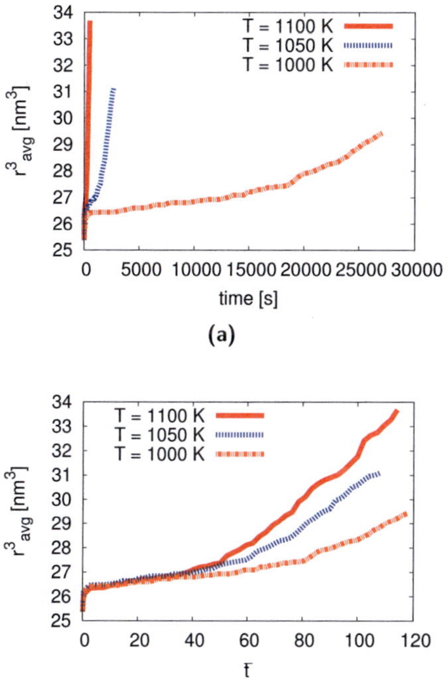

(b)

Figure 15.6: The effect of temperature on coarsening kinetics. The cubes of the average precipitate radii are plotted over time in (a) and over a nondimensionalized time in (b).

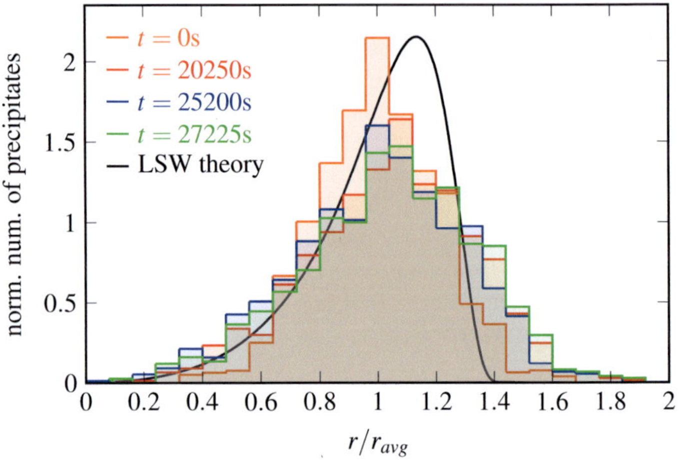

(a)

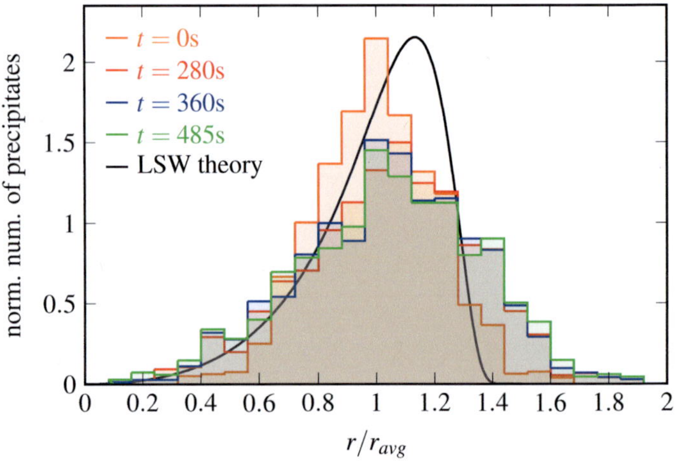

(b)

Figure 15.7: The effect of different inter-diffusivities and temperatures on the scaled size distributions. The histograms of the normalized numbers of precipitates are plotted over the normalized radii for three different simulation times. At $t = 0$ the domain is filled with a Gaussian distribution of particles. The predictions from the LSW theory are drawn as black lines. The distributions in red, blue and green correspond to the steady-state and are self-similar. Figure (a) shows the results for $T = 1000$ K and figure (b) for $T = 1100$ K.

15.3 Conclusions

A quantitative phase-field model combined with thermodynamic data from the CALPHAD method and diffusivity data from computational kinetics is used to study precipitate coarsening in the system Cu-Fe. Although the Gibbs energies from CALPHAD feature a miscibility gap, a description for the use with the grand potential model can be achieved by modeling the free energies as two independent parabolas in the vicinity of the equilibrium compositions. The coarsening kinetics from the simulations follow the LSW temporal power law in the long-time regime. The changing of temperature and hence diffusivity has a strong influence on the time-scale, at which the coarsening takes place. Nevertheless, the resulting distributions of the precipitates are statistically equivalent with a dynamic scaling behavior of the steady-state distributions for normalized numbers of precipitates.

In summary it can be stated that by the use of the phase-field method, diffusion driven processes of a more complex geometry can be treated, as it is possible with the one-dimensional models from computational kinetics tools (see section 5.2), such as DICTRA. Conversely, if accurate data from computational kinetics is available, it can be used for phase-field simulations of diffusion driven processes in order to raise their level of quantitativeness.

Part V

Conclusions and Outlook

16 Conclusions

The grand potential model has proven its ability to produce valuable research results for material processes such as solidification [66] or solid state transformations [63]. Nevertheless, quantitative findings can only be achieved if the simulations are based on accurate input data. The coupling of the phase-field model to thermodynamic databases is therefore of particular importance and plays a central role in this thesis. As a practical test of the presented coupling framework, it is applied for the simulation of actual material systems and the results are compared to well-established theories. For the solidification of Al-Si(-Mg), important influencing factors are identified, which are crucial for the resulting morphologies and the comparability to the analytical models.

16.1 CALPHAD coupling

As the grand potential model is explained in detail elsewhere [48, 7], it is addressed rather shortly in this dissertation. Instead, the focus is laid on the utilization of thermodynamic data provided by the CALPHAD method. The underlying idea of the coupling procedures is to change over from a phenomenological thermodynamic model designed for the calculation of phase diagrams to a phenomenological thermodynamic model, which is optimized for phase-field simulations. This topic also gets thematized in a publication from Choudhury et al. [56], where the approach of quadratic Taylor expansions is discussed explicitly. The intention of this thesis is to clearly categorize the different aspects connected to thermodynamic modeling and to distinguish between the choice of formulations and the methods of parametrization. Combining these different aspects leads to a collection of practices to cope with the characteristic features of diverse material systems. However, the presented list of coupling strategies is still not exhaustive, as for example the use of lookup tables is not taken into consideration. In the scope of this thesis, only the treatment of binary and ternary systems is covered and no systems of higher order are taken into account. Nevertheless, the presented simulation framework is in principle also suited for higher numbers of components.

For being able to derive and test the thermodynamic approximations in a comfortable and fast way, an infrastructure of preprocessing tools and interfaces has been developed for the phase-field software package PACE 3D from the research group of professor Nestler. As the IT infrastructures of other organizations can vary significantly, only the principal preprocessing framework but not its actual implementation gets addressed in this thesis. Despite all aid from software tools, a certain expertise is indispensable to derive appropriate thermodynamical descriptions. Due to the variety of different simulative tasks, no standard procedure can be recommended, which is suited best for every case. The choice of the coupling strategy always has to be made depending upon the material system and the specific objectives of the research study. Before the start of a simulation, it is recommended to check out the suitability of the particular approximation with basic setups. One can test, for instance in 1D domains, whether the model is capable of reproducing the relevant equilibrium concentrations and phase fractions according to the lever rule. In any case, one has to pay regard to the assumption of simulating close to equilibrium conditions. For high undercoolings, this condition is not fulfilled and the applicability of the grand potential model for such a task needs to be reconsidered.

Within the scope of this thesis, the utilization of CALPHAD databases is presented solely with reference to the grand potential model. Each phase-field model requires a specific set of input parameters and therefore different strategies for the incorporation of thermodynamic data are required. However, polynomial formulations and the methods of least squares and Taylor approximations are widely used for any kind of approximation task and thus can also be applied for the thermodynamic coupling to other phase-field models. An ongoing progress in computer technology will allow the use of more complex descriptions and hence enable higher degrees of accuracy. Nevertheless, it is a fundamental characteristic of models, that they never include all attributes of the real physical processes and just reproduce the most important factors of influence. It is thus justified and consequent to apply unsophisticated formulations in order to model the relevant thermodynamic properties of a material system.

16.2 Phase-field simulations

As a second topic of this thesis, the strategies for the utilization of thermodynamic data are applied for phase-field simulations of relevant phase transformation processes. In doing so, the thematic priority is on the solidification of aluminum-silicon, which gets studied by systematic variation of the input parameters. Different amounts of the ternary component magnesium modify the formation of microstructures, such as the transition from planar to cellular and dendritic growth fronts. The study of coarsening in the system iron-copper is an example for the application of diffusion data from computational kinetics, which significantly influences the resulting time-scale. An application of computational kinetics for the process of solidification was not an option, as no accurate mobility data is provided for the involved liquid phase. However, the availability of data from computational thermodynamics and kinetics is not the only condition for quantitative phase-field modeling. The simulation results for Al-Si show a strong dependence on the surface energy and the strength of its anisotropy, which emphasizes the demand for such kind of experimental data. Within the scope of this thesis, the solidification of Al-Si is studied in the hypo-eutectic regime on the Al-rich side of the phase-diagram. Therefore, only the primary solidification phase FCC and the liquid phase are involved in the phase transformation, whereas the stoichiometric diamond Si-phase is not part of the studies. The inclusion of the secondary diamond phase and the influence of its faceted anisotropy on the growth morphologies offers a lot of possibilities for future simulation series, for example on interdendritic eutectics [123]. The focus of this thesis is, however, on the essential growth modes of single phases, which are studied in basic setups to enable the comparability of the results.

The grand potential model is capable of producing results, which are in consistence with the established solidification theories from Mullins-Sekerka and Lipton-Glicksman-Kurz or the Lifshitz-Slyozov-Wagner law for coarsening. Such analytical solutions are an important basis of comparison for the phase-field model and its parametrizations. However, these theories bear on certain simplifications and cannot represent physical processes in their entirety. A good match between simulations and theory can only be achieved if the specific assumptions of the analytical solutions are fulfilled in the simulation setup, such as parabolic dendrite tips or idealized perturbations of sinusoidal shape. The phase-field model is able to describe more complex geometries than any of the analytical solutions and can generate realistic microstructures as obtained in experiments. Significant comparisons with real microstructures, however, necessitate big simulation domains in order to reduce the influence of the boundary conditions and to enable a statistical quantification of the morphologies. This is especially true for dendritic crystals, which require a fine grid for resolving their fractal structure. A more exact representation of the studied processes is possible if the simulations are carried out in 3D and also include additional physical phenomena, as for example the effect of fluid flow. Still, one should not underrate the meaningfulness of basic simulation setups. Studying only certain details of processes and excluding physical influencing factors, can help isolate the most relevant parameters for different types of microstructure formation. Some effects only appear through the interplay of several physical mechanisms and by the systematic combination and separation of influencing factors, simulations can help to identify such interdependencies. This flexibility in the choice of the setups is a notable advantage of simulations over real experiments.

17 Outlook

The work presented in this thesis belongs to the first generation of studies with the new and not fully optimized implementation of the grand potential model. Therefore, the results are not at the latest state of the art in terms of high performance computing. With the ongoing increase in computer power, simulations in three dimensions and large domains are about to become computable in a reasonable amount of time, which enables predictions of greater detail. Including further optimizations, the next generation of phase-field simulations is already running on leading supercomputers. The results of these studies are very promising, as they closely resemble experimental micrographs [55, 124]. The overall objective of designing and optimizing materials and their manufacturing processes by cost-efficient simulations is therefore within one's reach. However, the complete replacement of experiments is not a realistic scenario, because the search for new materials requires the strengths of different approaches and the synergy effects arising from their combination.

At the end of this thesis, I would like to draw a parallel between the current progress in microstructural simulation methods and the development of meteorological prediction techniques. These models have been constantly optimized during the last decades and are now able to prognosticate the complex processes of weather in a reliable way. Besides enough computational resources, long range forecasts require the indispensable data from numerous measuring stations. Analogously, microstructural simulation methods have the potential to generate valuable predictions about the evolution of multiphase and multi-component systems, if they include profound material parameters from experiments. Looking back on the achieved successes of the past years, I believe that quantitative phase-field modeling in combination with efficient high-performance computing offers bright prospects for tomorrow's material development.

Part VI

Appendix

A Deviation analysis for Gibbs energy approximations

In section 12.2.1 the difference $\Delta G(c) = {}^{id}G(c) - G_T(c)$ between the ideal part of the CALPHAD Gibbs energy description and its second order Taylor approximation is investigated. For being able to express $|\Delta G(c)|$ as a piecewise function in the interval $I_0 := [0, 0.5]$ for any c_0 inside of $\overline{I_0} := (0, 0.5]$, the algebraic sign of $\Delta G(c)$ in dependence of c_0 has to be known. For this purpose, the first derivative of $\Delta G(c)$ is analyzed inside of the interval $\overline{I_0}$. It is given as the sum of the derivatives of the individual terms:

$$\Delta G'(c) = {}^{id}G'(c) - G'_T(c). \tag{A.1}$$

Because $G_T(c)$ is the second-order Taylor approximation of ${}^{id}G(c)$ around c_0, the derivative $G'_T(c)$ is the first-order Taylor approximation of ${}^{id}G'(c)$ around that point. The second derivative of ${}^{id}G'(c)$ is ${}^{id}G'''(c)$ (given in equation 12.4) and for $c \in \overline{I_0}$ it is non-positive. From this property of its second derivative it follows, that ${}^{id}G'(c)$ is concave in $\overline{I_0}$. It has been proven [125, p. 489], that if a function is concave and differentiable inside of an interval, then it is bounded above by its Taylor approximation of first order inside of this interval. As this is the case for the function ${}^{id}G'(c)$ with $c, c_0 \in \overline{I_0}$, the inequation

$${}^{id}G'(c) \leq G'_T(c) \tag{A.2}$$

holds in this interval. The property, that $G'_T(c)$ is the tangent to ${}^{id}G'(c)$ at c_0, can also be seen in figure A.1. The derivative of $\Delta G(c)$ is thus non-positive for $c, c_0 \in \overline{I_0}$ and therefore $\Delta G(c)$ is monotonically decreasing for this case. Due to the monotonicity of the function and because $\Delta G(c)$ vanishes at c_0 it is thus justified to define the absolute value function inside of the interval $\overline{I_0}$ like in equation 12.6.

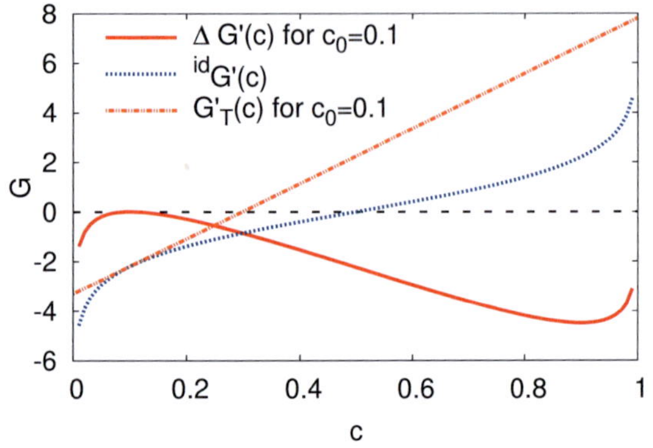

Figure A.1: The function $\Delta G'(c)$ is the derivative of the difference between the ideal part ${}^{id}G(c)$ and its parabolic approximation $G_T(c)$. This derivative function consists of the derivatives of both parts ${}^{id}G'(c)$ and $G_T'(c)$. The plot shows the approximation for $c_0 = 0.1$.

To provide a piecewise function of $|\Delta G(c)|$ also for the closed interval I_0, the Taylor approximation for $c = 0$

$$G_T(0) = {}^{id}G(c_0) - c_0 \cdot {}^{id}G'(c_0) + \frac{{}^{id}G''(c_0)}{2} c_0^2 \tag{A.3}$$

gets analyzed. This formula can be simplified to

$$G_T(0) = \frac{c_0}{2(1-c_0)} + \ln(1-c_0) \tag{A.4}$$

and the natural logarithm can be expressed by using a Taylor series around 1 as

$$G_T(0) = \frac{c_0}{2(1-c_0)} - \left(c_0 + \frac{c_0^2}{2} + \frac{c_0^3}{3} + \frac{c_0^4}{4} + \cdots\right) \tag{A.5}$$

$$= \frac{1}{1-c_0}\left(c_0^2 - \frac{c_0}{2}\right) - \left(\frac{c_0^2}{2} + \frac{c_0^3}{3} + \frac{c_0^4}{4} + \cdots\right). \tag{A.6}$$

With $1 - c_0 > 0$ and $c_0^2 - \frac{c_0}{2} \leq 0$ it follows from this equation, that

$$G_T(0) \leq 0 \tag{A.7}$$

for $c_0 \in \overline{I_0}$ and the difference $\Delta G(0)$ is non-negative for these cases. The inequation A.2 therefore also holds for $c \in I_0, c_0 \in \overline{I_0}$.

For the special case $c_0 = 0.5$ the algebraic sign of the difference $\Delta G_{c_0=0.5}(c)$ in dependence of concentration can also be determined. It is already proven from analyzing the derivative functions, that $\Delta G(c)$ is non-negative for every $c_0 \in \overline{I_0}$ including the case $c_0 = 0.5$. In the same way it follows from ${}^{id}G'''(c)$ being non-negative for $c \in \overline{I_1}$, that ${}^{id}G'(c)$ is convex in $\overline{I_1}$. Therefore in the convex region $c \in \overline{I_1}$ the inequation

$${}^{id}G'(c) \geq G'_{T,c_0=0.5}(c) \tag{A.8}$$

holds, because a differentiable function, which is convex inside of an interval, is greater or equal than any of its tangents inside of this interval. The function $\Delta G_{c_0=0.5}(c)$ is thus monotonically decreasing for $c \in \overline{I_0}$ and monotonically increasing for $c \in \overline{I_1}$. With $\Delta G_{c_0=0.5}(0.5) = 0$ it follows, that $\Delta G_{c_0=0.5}(c) \geq 0$ and $|\Delta G_{c_0=0.5}(c)|$ can be equated with $\Delta G_{c_0=0.5}(c)$.

B Deviation analysis for approximations of the phase concentrations

In addition to the analysis of the deviations arising from the use of parabolic expressions for the Gibbs energies in section 12.2.1, the approximated functions for the phase concentrations are analyzed in the following. For the ideal part of the Gibbs energies considered in the previous analysis, the analytical function for the chemical potential is given by the first derivative with respect to concentration in equation 12.2. It has to be mentioned, that the chemical potential is hereby a nondimensional quantity, just like the Gibbs energies used in this example. By inverting this expression, the analytical function $c_{id}(\mu)$ for the phase concentrations can be derived as

$$c_{id}(\mu) = \frac{1}{2}\left(\tanh\left(\frac{\mu}{2}\right) + 1\right). \tag{B.1}$$

An expression for the chemical potential can also be derived from the Taylor approximation $G_T(c)$ used in section 12.2.1 and is given by formula 11.6. Rearranging this expression leads to an approximated function for the phase concentrations $c_T(\mu)$ according to equation 11.6, which is itself the first order Taylor approximation of $c_{id}(\mu)$ around $\mu_0 = {}^{id}G'(c_0)$. A plot of $c_T(\mu)$ and $c_{id}(\mu)$ together with their difference $\Delta c(\mu) = c_{id}(\mu) - c_T(\mu)$ is given in figure B.1(a). It can be seen, that despite the root of $\Delta c(\mu)$ at μ_0 there can be a second root at $\mu_1 = {}^{id}G'(c_1)$. If a concentration c is used to calculate the analytical chemical potential $\mu = {}^{id}G'(c)$ and then the concentration is recalculated from μ with the approximated function $c_T(\mu)$, the result differs from the original concentration, despite for c_0 and c_1. A plot of such a recalculation for $c_0 = 0.25$ is given in figure B.1(b), including also a graph of the function $c_{id}(c) = c$. It can be seen, that $c_{id}(c)$ is the tangent to $c_T(c)$ at c_0 for such a plot over the concentration.

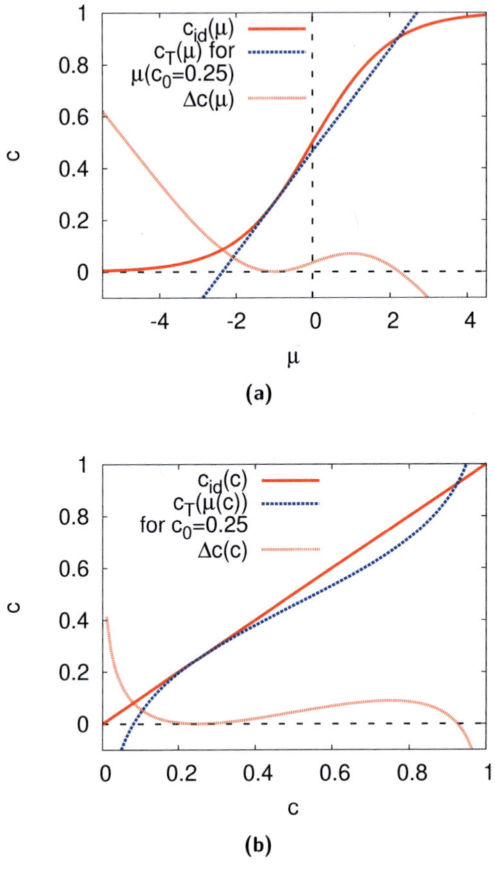

Figure B.1: (a) The analytical phase concentrations $c_{id}(\mu)$ are plotted over μ together with the approximated concentrations $c_T(\mu)$ (fitted around $\mu_0 = \ln(\frac{1}{3}) = \mu(c_0 = 0.25)$) and their difference $\Delta c(\mu)$. (b) The same functions as in (a) are plotted over the corresponding concentrations, calculated as $c = c_{id}(\mu)$.

According to equation 11.6, the second derivative of $c_T(c)$ is

$$c''_T({}^{id}G'(c)) = \frac{{}^{id}G'''(c)}{{}^{id}G''(c_0)} \tag{B.2}$$

and for $c, c_0 \in \overline{I_0}$ it is non-positive, such that $c_T(c)$ is concave. Because this entails, that the tangents $c_{id}(c)$ are not located below $c_T(c)$ for $c, c_0 \in \overline{I_0}$, the difference between the original and the recalculated composition

$$\Delta c(c) = c - c_T({}^{id}G'(c)) = c - c_0 - \frac{{}^{id}G'(c) - {}^{id}G'(c_0)}{{}^{id}G''(c_0)} \tag{B.3}$$

is non-negative in this case and can be used to express $|\Delta c(c)|$. In a similar manner as for the analysis of the Gibbs energy approximations the integral over the concentration

$$\int |\Delta c(c)|\, dc = \frac{c}{{}^{id}G''(c_0)} \left({}^{id}G'(c_0) + \frac{{}^{id}G''(c_0)}{2}(c - 2c_0) - {}^{id}G'(c) - \frac{\ln(c-1)}{c} \right) + C \tag{B.4}$$

can be used to quantify the error for $c \in I_0$ and $c_0 \in \overline{I_0}$ connected to the integration

$$\int_{c_0-\varepsilon}^{c_0+\varepsilon} |\Delta c(c)|\, dc = \frac{1}{{}^{id}G''(c_0)} \Big({}^{id}G'(c_0) 2\varepsilon + (c_0 - \varepsilon)^{id}G'(c_0 - \varepsilon) - (c_0 + \varepsilon)^{id}G'(c_0 + \varepsilon) + \ln(c_0 - 1 - \varepsilon) - \ln(c_0 - 1 + \varepsilon) \Big) \tag{B.5}$$

over an interval around c_0 of width 2ε. Because $c_T(0)$ is not defined, $0 \leq \varepsilon < \min(c_0, 0.5 - c_0)$ has to be satisfied hereby. For the case $c_0 = 0.5$ the second derivative of the approximation $c''_T(c)$ is non-negative for $c \in \overline{I_1}$, such that $c_T(c)$ is convex. Because this entails, that the tangents $c_{id}(c)$ are not located above $c_T(c)$ in this case, the difference between the original and the recalculated composition is non-positive. As the difference is non-negative for $c \in \overline{I_0}$, the integration around $c_0 = 0.5$ thus has to be done like

$$\int_{0.5-\varepsilon}^{0.5+\varepsilon} |\Delta c(c)|\, dc = \int_{0.5-\varepsilon}^{0.5} \Delta c(c)\, dc - \int_{0.5}^{0.5+\varepsilon} \Delta c(c)\, dc \tag{B.6}$$

and the definite integral results as

$$\int_{0.5-\varepsilon}^{0.5+\varepsilon} |\Delta c(c)|\, dc = \varepsilon\, \text{artanh}(2\varepsilon) + \frac{1}{4} \ln\left(1 - 4\varepsilon^2\right) - \varepsilon^2. \tag{B.7}$$

A plot of the result for this integration around $c_0 = 0.5$ divided by 2ε is shown in figure B.2 together with the results of equation B.5 divided by 2ε plotted for varying c_0. Similarly to the deviations of the approximated Gibbs energies (plotted in figure 12.3(a)) the deviation of c_T is the smallest close to $c_0 = 0.5$ and the largest close to the terminal compositions. It is worth mentioning, that the analytical phase concentrations can also be derived if in addition to the ideal part of the CALPHAD formulation the reference part is taken into account. The deviations in dependence of c_0 are similar for these cases. For more sophisticated Gibbs energy expressions the inverse functions can not be derived that easily.

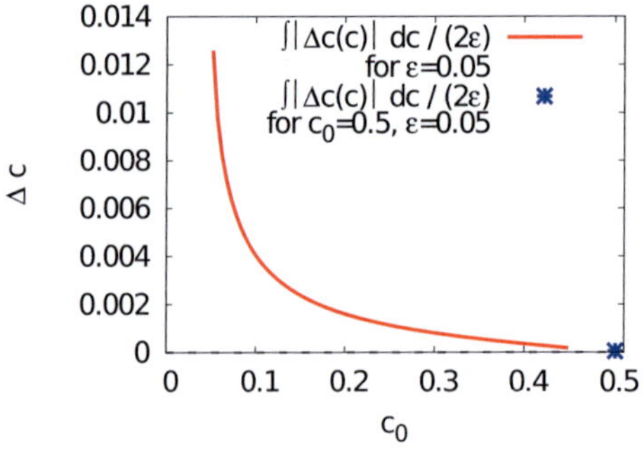

Figure B.2: The integral of the difference between the analytical phase concentrations and their Taylor approximations for varying $c_0 \in \overline{I_0}$. The integration is done in an interval of width 2ε around c_0 and divided by 2ε.

C Proof of the invertibility of the chemical potential functions of the Al-Si system

The fitting of functions for the phase concentrations in dependence of the chemical potentials rests upon the invertibility of the functions for the chemical potential in dependence of the concentrations. For the FCC and liquid phase from the system Al-Si, which is treated in section 14.1, this invertibility is proven exemplarily in the following. As the fitting is based on molar Gibbs energies, the chemical potentials $\mu_m = \partial G_m / \partial c$ are also molar quantities, with c as the concentration of aluminum. In the formulation used in the dataset of Feufel et al., the excess part of the FCC phase is modeled with a Redlich-Kister series of zeroth order. By applying equations 5.8 to 5.11, the function for $\mu_m^{FCC}(T,c)$ can be derived as

$$\mu_m^{FCC}(T,c) = {}^{\circ}G_{Al}^{FCC}(T) - {}^{\circ}G_{Si}^{FCC}(T) + RT\ln\left(\frac{c}{1-c}\right) - {}^{0}L_{Al,Si}^{FCC}(T)(2c-1) \tag{C.1}$$

and the derivative of μ_m^{FCC} with respect to c is given as:

$$\frac{\partial \mu_m^{FCC}(T,c)}{\partial c} = \frac{RT}{c(1-c)} - 2\cdot {}^{0}L_{Al,Si}^{FCC}(T). \tag{C.2}$$

By inserting the expression for ${}^{0}L_{Al,Si}^{FCC}(T)$ from Table 14.1, one gets:

$$\frac{\partial \mu_m^{FCC}(T,c)}{\partial c} = \frac{RT}{c(1-c)} + 22680.2\ \mathrm{J/mol} + T \cdot 2.46788\ \mathrm{J/molK}. \qquad \text{(C.3)}$$

From the definition of a temperature in the Kelvin scale being positive and the definition $0 \leq c \leq 1$, it follows that $\partial \mu_m^{FCC}(T,c)/\partial c$ is positive for any temperature and concentration. This proves the monotonicity of $\mu_m^{FCC}(T,c)$ and the existence of the inverse function $c^{FCC}(T,\mu_m)$.

The excess part of the liquid phase is described by a Redlich-Kister series of second order. A formula for μ_m^{liq} can analytically be determined just in the same way as for the FCC phase. This leads to a formula dependent on c of degree three. Instead of using the analytical expression, the function for μ_m^{liq} is derived by fitting. For doing this, the values of the sum of the reference and excess part given by equations 5.9 and 5.11 are calculated at the lower and upper temperature limit for 100 different concentrations inside of the concentration range between $c = 0.9$ and $c = 0.995$. With the choice of a quadratic expression for the sum of the reference and excess part, the coefficients are evaluated by curve fitting. By adding the contribution from the ideal part, formulas for the total chemical potential of the liquid phase are obtained. For the two constant temperatures T^*, the terms are of the following kind:

$$\mu_m^{liq}(T^*,c) = b_0 + b_1 \cdot c + b_2 \cdot c^2 + RT^* \ln\left(\frac{c}{1-c}\right). \qquad \text{(C.4)}$$

To prove the strict monotonicity of $\mu_m^{liq}(T^*,c)$, one can take the derivative of the derived polynomial from equation C.4 with respect to c and get:

$$\frac{\partial \mu_m^{liq}(T^*,c)}{\partial c} = b_1 + 2b_2 \cdot c + \frac{RT^*}{c(1-c)}. \qquad \text{(C.5)}$$

One can now assume that the derivative is positive for all concentrations at a constant temperature. This is true if the following inequation is fulfilled:

$$-\frac{RT^*}{c(1-c)} \leq b_1 + 2b_2 \cdot c. \tag{C.6}$$

Due to the definition of temperature and concentration, the left side of the equation is negative in all cases. Therefore it is sufficient if the right side of the equation is positive, which is true if the following inequation is fulfilled:

$$-2b_2 \cdot c \leq b_1. \tag{C.7}$$

This leads to the conditions

$$c \leq -\frac{b_1}{2b_2} \text{ if } b_2 < 0, \tag{C.8}$$

$$-c \leq \frac{b_1}{2b_2} \text{ if } b_2 > 0. \tag{C.9}$$

From the isothermal fit at $T^* = 875$ K, the coefficients result as $b_1 = 85608.565$ J/mol and $b_2 = -34560.26$ J/mol and in consequence $c \leq 1.239$ is fulfilled over the whole concentration range. Hereby the evidence for the monotonicity of $\mu_m^{liq}(T^*, c)$ and hence the existence of the inverse function $c^{liq}(T^*, \mu_m)$ is given for the specific temperature.

Bibliography

[1] B. Nestler and A. Choudhury, "Phase-field modeling of multi-component systems," *Current opinion in solid state and Materials Science*, vol. 15, no. 3, pp. 93–105, 2011.

[2] L. Kaufman and H. Bernstein, "Computer calculation of phase diagrams," *Academic Press, New York*, 1970.

[3] G. Cole and A. Sherman, "Light weight materials for automotive applications," *Materials characterization*, vol. 35, no. 1, pp. 3–9, 1995.

[4] A. Morita, "Aluminum alloys for automobile applications," in *ICAA-6: 6 th International Conference on Aluminium Alloys, Toyohashi, Japan, 5-10 Japan 1998*, 1998, pp. 25–32.

[5] W. Miller, L. Zhuang, J. Bottema, A. Wittebrood, P. De Smet, A. Haszler, and A. Vieregge, "Recent development in aluminium alloys for the automotive industry," *Materials Science and Engineering: A*, vol. 280, no. 1, pp. 37–49, 2000.

[6] P. Spencer, "A brief history of calphad," *Calphad*, vol. 32, no. 1, pp. 1–8, 2008.

[7] A. Choudhury and B. Nestler, "Grand-potential formulation for multicomponent phase transformations combined with thin-interface asymptotics of the double obstacle potential," *Phys. Rev. E*, vol. 85, p. 021602, 2012.

[8] W. W. Mullins and R. Sekerka, "Stability of a planar interface during solidification of a dilute binary alloy," *Journal of applied physics*, vol. 35, no. 2, pp. 444–451, 1964.

[9] J. Lipton, M. Glicksman, and W. Kurz, "Dendritic growth into undercooled alloy metals," *Materials Science and Engineering*, vol. 65, no. 1, pp. 57–63, 1984.

[10] J. W. Gibbs, *The Scientific Papers, Vol I: Thermodynamics.* Dover Publications, 1961.

[11] D. A. Porter and K. E. Easterling, *Phase Transformations in Metals and Alloys, (Revised Reprint).* CRC press, 1992.

[12] H. L. Lukas, S. G. Fries, and B. Sundman, *Computational thermodynamics: the CALPHAD method.* Cambridge University Press, 2007.

[13] K. Hack, *The SGTE casebook: thermodynamics at work.* Elsevier, 2008.

[14] J.-O. Andersson, T. Helander, L. Höglund, P. Shi, and B. Sundman, "Thermo-calc & dictra, computational tools for materials science," *Calphad*, vol. 26, no. 2, pp. 273–312, 2002.

[15] S.-L. Chen, S. Daniel, F. Zhang, Y. Chang, X.-Y. Yan, F.-Y. Xie, R. Schmid-Fetzer, and W. Oates, "The pandat software package and its applications," *Calphad*, vol. 26, no. 2, pp. 175–188, 2002.

[16] C. Bale, E. Bélisle, P. Chartrand, S. Decterov, G. Eriksson, K. Hack, I.-H. Jung, Y.-B. Kang, J. Melançon, A. Pelton *et al.*, "Factsage thermochemical software and databases—recent developments," *Calphad*, vol. 33, no. 2, pp. 295–311, 2009.

[17] R. Davies, A. Dinsdale, J. Gisby, J. Robinson, and S. Martin, "Mtdata-thermodynamic and phase equilibrium software from the national physical laboratory," *Calphad*, vol. 26, no. 2, pp. 229–271, 2002.

[18] B. Sundman, U. R. Kattner, M. Palumbo, and S. G. Fries, "Opencalphad-a free thermodynamic software," *Integrating Materials and Manufacturing Innovation*, no. 1, p. 1, 2015.

[19] L. K. (Ed.), "CALPHAD — Computer Coupling of Phase Diagrams and Thermochemistry, Pergamon, Oxford," 1977 onward.

[20] A. T. Dinsdale, "CALPHAD 15," 1991, 317-425.

[21] J.-C. Zhao, *Methods for phase diagram determination.* elsevier, 2011.

[22] M. Hillert, "The compound energy formalism," *Journal of Alloys and Compounds*, vol. 320, no. 2, pp. 161–176, 2001.

[23] M. Hillert and L.-I. Staffanson, *Acta Chem. Scand.*, vol. 24, pp. 3618–3626, 1970.

[24] B. Sundman and J. Ågren, "A regular solution model for phases with several components and sublattices, suitable for computer applications," *Journal of physics and chemistry of solids*, vol. 42, no. 4, pp. 297–301, 1981.

[25] O. Redlich and A. Kister, "Algebraic representation of thermodynamic properties and the classification of solutions," *Industrial & Engineering Chemistry*, vol. 40, no. 2, pp. 345–348, 1948.

[26] Y. Muggianu, M. Gambino, and L. Bross, "Comparison between calculated and measured thermodynamic data of liquid (ag, au, cu)-sn-zn alloys," *J. Chim. Phys*, vol. 72, pp. 85–91, 1975.

[27] M. Hillert, "Empirical methods of predicting and representing thermodynamic properties of ternary solution phases," *Calphad*, vol. 4, no. 1, pp. 1–12, 1980.

[28] L. Kaufman, "Coupled phase diagrams and thermochemical data for transition metal binary systems-vi," *Calphad*, vol. 3, no. 1, pp. 45–76, 1979. [Online]. Available: http://www.sciencedirect.com/science/article/pii/0364591679900208

[29] H. Feufel, T. Gödecke, H. L. Lukas, and F. Sommer, "Investigation of the al-mg-si system by experiments and thermodynamic calculations," *Journal of Alloys and compounds*, vol. 247, no. 1, pp. 31–42, 1997.

[30] J. Gröbner, H. L. Lukas, and F. Aldinger, "Thermodynamic calculation of the ternary system al-si-c," *Calphad*, vol. 20, no. 2, pp. 247–254, 1996.

[31] M. Hillert and M. Jarl, "A model for alloying in ferromagnetic metals," *Calphad*, vol. 2, no. 3, pp. 227–238, 1978.

[32] A. Borgenstam, L. Höglund, J. Ågren, and A. Engström, "Dictra, a tool for simulation of diffusional transformations in alloys," *Journal of phase equilibria*, vol. 21, no. 3, pp. 269–280, 2000.

[33] J.-O. Andersson and J. Ågren, "Models for numerical treatment of multicomponent diffusion in simple phases," *Journal of Applied Physics*, vol. 72, no. 4, pp. 1350–1355, 1992.

[34] W. Kurz and D. J. Fisher, *Fundamentals of Solidification*, 4th ed. Trans Tech Publications, 1998.

[35] G. Ivantsov, "Temperature around a spheroidal, cylindrical and acicular crystal growing in a supercooled melt," in *Dokl. Akad. Nauk*, vol. 58, 1947, pp. 567–569.

[36] G. Horvay and J. Cahn, "Dendritic and spheroidal growth," *Acta Metallurgica*, vol. 9, no. 7, pp. 695–705, 1961.

[37] J. Langer and J. Müller-Krumbhaar, "Stability effects in dendritic crystal growth," *Journal of Crystal Growth*, vol. 42, pp. 11–14, 1977.

[38] A. Barbieri and J. Langer, "Predictions of dendritic growth rates in the linearized solvability theory," *Physical Review A*, vol. 39, no. 10, p. 5314, 1989.

[39] V. L. Ginzburg and L. D. Landau, *Sov. Phys. JETP*, vol. 20, p. 1064, 1950.

[40] J. W. Cahn and J. E. Hilliard, "Free energy of a nonuniform system. i. interfacial free energy," *The Journal of chemical physics*, vol. 28, no. 2, pp. 258–267, 1958.

[41] R. Kobayashi, "Modeling and numerical simulations of dendritic crystal growth," *Physica D: Nonlinear Phenomena*, vol. 63, no. 3, pp. 410–423, 1993.

[42] I. Steinbach, F. Pezzolla, B. Nestler, M. Seeßelberg, R. Prieler, G. Schmitz, and J. Rezende, "A phase field concept for multiphase systems," *Physica D: Nonlinear Phenomena*, vol. 94, no. 3, pp. 135–147, 1996.

[43] B. Nestler, H. Garcke, and B. Stinner, "Multicomponent alloy solidification: Phase-field modeling and simulations," *Physical Review E*, vol. 71, no. 4, p. 041609, 2005.

[44] S. G. Kim, W. T. Kim, and T. Suzuki, "Interfacial compositions of solid and liquid in a phase-field model with finite interface thickness for isothermal solidification in binary alloys," *Physical Review E*, vol. 58, no. 3, p. 3316, 1998.

[45] J. Tiaden, B. Nestler, H. Diepers, and I. Steinbach, "The multiphase-field model with an integrated concept for modelling solute diffusion," *Physica D: Nonlinear Phenomena*, vol. 115, no. 1, pp. 73–86, 1998.

[46] J. Eiken, B. Böttger, and I. Steinbach, "Multiphase-field approach for multicomponent alloys with extrapolation scheme for numerical application," *Physical Review E*, vol. 73, no. 6, p. 066122, 2006.

[47] M. Plapp, "Unified derivation of phase-field models for alloy solidification from a grand-potential functional," *Physical Review E*, vol. 84, no. 3, p. 031601, 2011.

[48] A. N. Choudhury, *Quantitative phase-field model for phase transformations in multi-component alloys*, ser. Schriftenreihe des Instituts für Angewandte Materialien, Karlsruher Institut für Technologie, 21. KIT Scientific Publishing, Karlsruhe, 2013.

[49] B. Echebarria, R. Folch, A. Karma, and M. Plapp, "Quantitative phase-field model of alloy solidification," *Physical Review E*, vol. 70, no. 6, p. 061604, 2004.

[50] N. Provatas and K. Elder, *Phase-field methods in materials science and engineering*. John Wiley & Sons, 2010.

[51] J. Heulens, B. Blanpain, and N. Moelans, "Phase-field analysis of a ternary two-phase diffusion couple with multiple analytical solutions," *Acta Materialia*, vol. 59, no. 10, pp. 3946–3954, 2011.

[52] A. Karma and W.-J. Rappel, "Phase-field method for computationally efficient modeling of solidification with arbitrary interface kinetics," *Physical Review E*, vol. 53, no. 4, p. R3017, 1996.

[53] A. Karma, "Phase-field formulation for quantitative modeling of alloy solidification," *Physical Review Letters*, vol. 87, no. 11, p. 115701, 2001.

[54] W. J. Boettinger and J. A. Warren, "Simulation of the cell to plane front transition during directional solidification at high velocity," *Journal of crystal growth*, vol. 200, no. 3, pp. 583–591, 1999.

[55] J. Hötzer, M. Jainta, P. Steinmetz, B. Nestler, A. Dennstedt, A. Genau, M. Bauer, H. Köstler, and U. Rüde, "Large scale phase-field simulations of directional ternary eutectic solidification," *Acta Materialia*, vol. 93, pp. 194–204, 2015.

[56] A. Choudhury, M. Kellner, and B. Nestler, "A method for coupling the phase-field model based on a grand-potential formalism to thermodynamic databases," *Current Opinion in Solid State and Materials Science*, vol. 19, no. 5, pp. 287–300, 2015.

[57] U. Grafe, B. Böttger, J. Tiaden, and S. Fries, "Coupling of multicomponent thermodynamic databases to a phase field model: application to solidification and solid state transformations of superalloys," *Scripta Materialia*, vol. 42, no. 12, pp. 1179–1186, 2000.

[58] J. Zhu, Z. Liu, V. Vaithyanathan, and L. Chen, "Linking phase-field model to calphad: application to precipitate shape evolution in ni-base alloys," *Scripta Materialia*, vol. 46, no. 5, pp. 401–406, 2002.

[59] R. Siquieri, E. Doernberg, H. Emmerich, and R. Schmid-Fetzer, "Phase-field simulation of peritectic solidification closely coupled with directional solidification experiments in an al–36 wt% ni alloy," *Journal of Physics: Condensed Matter*, vol. 21, no. 46, p. 464112, 2009.

[60] R. Qin, E. Wallach, and R. Thomson, "A phase-field model for the solidification of multicomponent and multiphase alloys," *Journal of crystal growth*, vol. 279, no. 1, pp. 163–169, 2005.

[61] H. Kobayashi, M. Ode, S. G. Kim, W. T. Kim, and T. Suzuki, "Phase-field model for solidification of ternary alloys coupled with thermodynamic database," *Scripta Materialia*, vol. 48, no. 6, pp. 689–694, 2003.

[62] L. Zhang, M. Stratmann, Y. Du, B. Sundman, and I. Steinbach, "Incorporating the calphad sublattice approach of ordering into the phase-field model with finite interface dissipation," *Acta Materialia*, vol. 88, pp. 156–169, 2015.

[63] K. Ankit, A. Choudhury, C. Qin, S. Schulz, M. McDaniel, and B. Nestler, "Theoretical and numerical study of lamellar eutectoid growth influenced by volume diffusion," *Acta Materialia*, vol. 61, no. 11, pp. 4245–4253, 2013.

[64] R. Folch and M. Plapp, "Quantitative phase-field modeling of two-phase growth," *Physical Review E*, vol. 72, no. 1, p. 011602, 2005.

[65] B. Stinner, B. Nestler, and H. Garcke, "A diffuse interface model for alloys with multiple components and phases," *SIAM Journal on Applied Mathematics*, vol. 64, no. 3, pp. 775–799, 2004.

[66] E. Wesner, A. Choudhury, A. August, M. Berghoff, and B. Nestler, "A phase-field study of large-scale dendrite fragmentation in al–cu," *Journal of Crystal Growth*, vol. 359, pp. 107–121, 2012.

[67] A. Choudhury, K. Reuther, E. Wesner, A. August, B. Nestler, and M. Rettenmayr, "Comparison of phase-field and cellular automaton models for dendritic solidification in al–cu alloy," *Computational Materials Science*, vol. 55, pp. 263–268, 2012.

[68] D. Molnar, R. Mukherjee, A. Choudhury, A. Mora, P. Binkele, M. Selzer, B. Nestler, and S. Schmauder, "Multiscale simulations on the coarsening of cu-rich precipitates in α-fe using kinetic monte carlo, molecular dynamics and phase-field simulations," *Acta Materialia*, vol. 60, no. 20, pp. 6961–6971, 2012.

[69] M. Kist, A. Choudhury, and B. Nestler, "Oscillative surface morphology in peritectic nial using phase-field modeling," *International Journal of Materials Research*, vol. 104, no. 11, pp. 1096–1107, 2013.

[70] A. Choudhury, M. Geeta, and B. Nestler, "Influence of solid-solid interface anisotropy on three-phase eutectic growth during directional solidification," *EPL (Europhysics Letters)*, vol. 101, no. 2, p. 26001, 2013.

[71] K. Ankit, R. Mukherjee, T. Mittnacht, and B. Nestler, "Deviations from cooperative growth mode during eutectoid transformation: Insights from a phase-field approach," *Acta Materialia*, vol. 81, pp. 204–210, 2014.

[72] P. Gustafson, "A thermodynamic evaluation of the fe–c system," *Scand. J. Metall.*, vol. 14, no. 5, pp. 259–267, 1985.

[73] S. Hu, J. Murray, H. Weiland, Z. Liu, and L. Chen, "Thermodynamic description and growth kinetics of stoichiometric precipitates in the phase-field approach," *Calphad*, vol. 31, no. 2, pp. 303–312, 2007.

[74] I. Kornilov *et al.*, "Phase Diagram of the Cr-NiAl-System (in Russian language)," *Izv. Sekt. Fiz.-Khim. Anal.,22, (1993) 111-116*, 1953.

[75] Y. A. Bagaryatskiy, "The cr corner of the cr-ni-al system and the cr-nial pseudobinary section," *Zh. Neorg. Khim*, vol. 3, pp. 722–728, 1958.

[76] P. Rogl, *Ternary alloys.* Weinheim: VCH, 1991, vol. 4: Al-Cd-Ce to Al-Cu-Ru.

[77] T. Velikanova, K. Korniyenko, and V. Sidorko, "Aluminium–chromium–nickel," 2004.

[78] W. Huang and Y. Chang, "Thermodynamic properties of the ni–al–cr system," *Intermetallics*, vol. 7, no. 8, pp. 863–874, 1999.

[79] N. Dupin, I. Ansara, and B. Sundman, "Thermodynamic re-assessment of the ternary system al-cr-ni," *Calphad*, vol. 25, no. 2, pp. 279–298, 2001.

[80] A. Kusoffsky, N. Dupin, and B. Sundman, "On the compound energy formalism applied to fcc ordering." *Calphad*, vol. 25, no. 4, pp. 549–565, 2001.

[81] T. Sasaki and Y. Kanada, "Practically fast multiple-precision evaluation of log (x)," *Journal of information processing*, vol. 5, no. 4, pp. 247–250, 1982.

[82] J. A. Warren and W. J. Boettinger, "Prediction of dendritic growth and microsegregation patterns in a binary alloy using the phase-field method," *Acta Metallurgica et Materialia*, vol. 43, no. 2, pp. 689–703, 1995.

[83] A. Karma and W.-J. Rappel, "Quantitative phase-field modeling of dendritic growth in two and three dimensions," *Physical review E*, vol. 57, no. 4, p. 4323, 1998.

[84] Y. Zhang, W. Chengzhi, L. Dianzhong, and L. Yiyi, "Phase field modeling of dendrite growth," *Acta Metallurgica Sinica (English Letters)*, vol. 22, no. 3, pp. 197–201, 2009.

[85] Y. Zhao and H. Hou, "Simulation for microstructure evolution of al-si alloys in solidification process," vol. 29, no. 1, p. 210, 2006.

[86] K. Ohsasa, T. Katsumi, R. Sugawara, and Y. Natsume, "Fractal characteristics of dendrite in aluminum alloys," vol. 33, no. 1, p. 012108, 2012.

[87] R. Zhang, M. Li, and J. Allison, "Phase-field study for the influence of solute interactions on solidification process in multicomponent alloys," *Computational Materials Science*, vol. 47, no. 3, pp. 832–838, 2010.

[88] Z. Bi and R. F. Sekerka, "Phase field modeling of shallow cells during directional solidification of a binary alloy," *Journal of crystal growth*, vol. 237, pp. 138–143, 2002.

[89] C. Lan, C. Shih, and M. Lee, "Quantitative phase field simulation of deep cells in directional solidification of an alloy," *Acta materialia*, vol. 53, no. 8, pp. 2285–2294, 2005.

[90] W. Losert, D. Stillman, H. Cummins, P. Kopczyński, W.-J. Rappel, and A. Karma, "Selection of doublet cellular patterns in directional solidification through spatially periodic perturbations," *Physical Review E*, vol. 58, no. 6, p. 7492, 1998.

[91] Z. Wang, J. Li, J. Wang, and Y. Zhou, "Phase field modeling the selection mechanism of primary dendritic spacing in directional solidification," *Acta Materialia*, vol. 60, no. 5, pp. 1957–1964, 2012.

[92] M. Amoorezaei, S. Gurevich, and N. Provatas, "Orientation selection in solidification patterning," *Acta Materialia*, vol. 60, no. 2, pp. 657–663, 2012.

[93] I. Steinbach, "Effect of interface anisotropy on spacing selection in constrained dendrite growth," *Acta Materialia*, vol. 56, no. 18, pp. 4965–4971, 2008.

[94] H.-J. Diepers, D. Ma, and I. Steinbach, "History effects during the selection of primary dendrite spacing. comparison of phase-field simulations with experimental observations," *Journal of crystal growth*, vol. 237, pp. 149–153, 2002.

[95] M. Amoorezaei, S. Gurevich, and N. Provatas, "Spacing characterization in al–cu alloys directionally solidified under transient growth conditions," *Acta Materialia*, vol. 58, no. 18, pp. 6115–6124, 2010.

[96] A. Badillo and C. Beckermann, "Phase-field simulation of the columnar-to-equiaxed transition in alloy solidification," *Acta Materialia*, vol. 54, no. 8, pp. 2015–2026, 2006.

[97] P. Kopczyński, W.-J. Rappel, and A. Karma, "Critical role of crystalline anisotropy in the stability of cellular array structures in directional solidification," *Physical review letters*, vol. 77, no. 16, p. 3387, 1996.

[98] M. Plapp and M. Dejmek, "Stability of hexagonal solidification patterns," *EPL (Europhysics Letters)*, vol. 65, no. 2, p. 276, 2004.

[99] Y. Ma and M. Plapp, "Phase-field simulations and geometrical characterization of cellular solidification fronts," *Journal of Crystal Growth*, vol. 385, pp. 140–147, 2014.

[100] D. Coates, S. Subramanian, and G. Purdy, "Solid-liquid interface stability during solidification of dilute ternary alloys," *TRANS MET SOC AIME*, vol. 242, no. 5, pp. 800–809, 1968.

[101] O. Hunziker, "Theory of plane front and dendritic growth in multicomponent alloys," *Acta materialia*, vol. 49, no. 20, pp. 4191–4203, 2001.

[102] S. Liu, R. Napolitano, and R. Trivedi, "Measurement of anisotropy of crystal-melt interfacial energy for a binary al–cu alloy," *Acta Materialia*, vol. 49, no. 20, pp. 4271–4276, 2001.

[103] R. Napolitano, S. Liu, and R. Trivedi, "Experimental measurement of anisotropy in crystal-melt interfacial energy," *Interface Science*, vol. 10, no. 2-3, pp. 217–232, 2002.

[104] M. Sabouri-Ghomi, N. Provatas, and M. Grant, "Solidification of a supercooled liquid in a narrow channel," *Physical review letters*, vol. 86, no. 22, p. 5084, 2001.

[105] Z. Wang, J. Wang, J. Li, G. Yang, and Y. Zhou, "Quantitative investigation of cellular growth in directional solidification by phase-field simulation," *Physical Review E*, vol. 84, no. 4, p. 041604, 2011.

[106] R. Zhang, T. Jing, W. Jie, and B. Liu, "Phase-field simulation of solidification in multicomponent alloys coupled with thermodynamic and diffusion mobility databases," *Acta materialia*, vol. 54, no. 8, pp. 2235–2239, 2006.

[107] F. Cherne Iii and P. Deymier, "Calculation of the transport properties of liquid aluminum with equilibrium and non-equilibrium molecular dynamics," *Scripta materialia*, vol. 45, no. 8, pp. 985–991, 2001.

[108] Y. Du, Y. Chang, B. Huang, W. Gong, Z. Jin, H. Xu, Z. Yuan, Y. Liu, Y. He, and F.-Y. Xie, "Diffusion coefficients of some solutes in fcc and liquid al: critical evaluation and correlation," *Materials Science and Engineering: A*, vol. 363, no. 1, pp. 140–151, 2003.

[109] S. Gurevich, A. Karma, M. Plapp, and R. Trivedi, "Phase-field study of three-dimensional steady-state growth shapes in directional solidification," *Physical Review E*, vol. 81, no. 1, p. 011603, 2010.

[110] M. Li and G. Yang, "Growth morphologies of a binary alloy with low anisotropy in directional solidification," *Acta Metallurgica Sinica (English Letters)*, vol. 20, no. 4, pp. 258–264, 2007.

[111] R. Mukherjee, A. Choudhury, and B. Nestler, "Composition pathway in fe–cu–ni alloy during coarsening," *Modelling and Simulation in Materials Science and Engineering*, vol. 21, no. 7, p. 075012, 2013.

[112] I. M. Lifshitz and V. V. Slyozov, "The kinetics of precipitation from supersaturated solid solutions," *Journal of Physics and Chemistry of Solids*, vol. 19, no. 1, pp. 35–50, 1961.

[113] C. Wagner, "Theorie der alterung von niederschlägen durch umlösen (ostwald-reifung)," *Zeitschrift für Elektrochemie, Berichte der Bunsengesellschaft für physikalische Chemie*, vol. 65, no. 7-8, pp. 581–591, 1961.

[114] C. Servant, B. Sundman, and O. Lyon, "Thermodynamic assessment of the cu-fe-ni system," *Calphad*, vol. 25, no. 1, pp. 79–95, 2001.

[115] K. G. F. Janssens, D. Raabe, E. Kozeschnik, M. A. Miodownik, and B. Nestler, *Computational materials engineering: an introduction to microstructure evolution.* Academic Press, 2010.

[116] B. Sonderegger and E. Kozeschnik, "Generalized nearest-neighbor broken-bond analysis of randomly oriented coherent interfaces in multicomponent fcc and bcc structures," *Metallurgical and Materials Transactions A*, vol. 40, no. 3, pp. 499–510, 2009.

[117] H. Mehrer, *Landolt-Börnstein, New Series, Vol. III/26.* Berlin: Springer, 1990.

[118] B. Jönsson, "On ferromagnetic ordering and lattice diffusion: a simple model," *Zeitschrift für Metallkunde*, vol. 83, no. 5, pp. 349–355, 1992.

[119] B. Jönsson, "Assessment of the mobilities of cr, fe and ni in bcc cr-fe-ni alloys," *ISIJ international*, vol. 35, no. 11, pp. 1415–1421, 1995.

[120] P. W. Voorhees, "The theory of ostwald ripening," *Journal of Statistical Physics*, vol. 38, no. 1-2, pp. 231–252, 1985.

[121] C. H. Kang and D. N. Yoon, "Coarsening of cobalt grains dispersed in liquid copper matrix," *Metallurgical Transactions A*, vol. 12, no. 1, pp. 65–69, 1981.

[122] Y. Masuda, R. Watanabe, and G. Kuczynski, "Sintering processes," *Materials Science Research (Plenum, New York, 1979)*, vol. 13, 1980.

[123] R. Grugel and W. Kurz, "Growth of interdendritic eutectic in directionally solidified al-si alloys," *Metallurgical Transactions A*, vol. 18, no. 13, pp. 1137–1142, 1987.

[124] J. Hötzer, P. Steinmetz, M. Jainta, S. Schulz, M. Kellner, B. Nestler, A. Genau, A. Dennstedt, M. Bauer, H. Köstler *et al.*, "Phase-field simulations of spiral growth during directional ternary eutectic solidification," *Acta Materialia*, vol. 106, pp. 249–259, 2016.

[125] H. R. Varian and W. Norton, *Microeconomic analysis*. Norton New York, 1992, vol. 2.

Schriftenreihe des Instituts für Angewandte Materialien

ISSN 2192-9963

Die Bände sind unter www.ksp.kit.edu als PDF frei verfügbar oder als Druckausgabe bestellbar.

Band 1 Prachai Norajitra
Divertor Development for a Future Fusion Power Plant. 2011
ISBN 978-3-86644-738-7

Band 2 Jürgen Prokop
Entwicklung von Spritzgießsonderverfahren zur Herstellung von Mikrobauteilen durch galvanische Replikation. 2011
ISBN 978-3-86644-755-4

Band 3 Theo Fett
New contributions to R-curves and bridging stresses – Applications of weight functions. 2012
ISBN 978-3-86644-836-0

Band 4 Jérôme Acker
Einfluss des Alkali/Niob-Verhältnisses und der Kupferdotierung auf das Sinterverhalten, die Strukturbildung und die Mikrostruktur von bleifreier Piezokeramik $(K_{0,5}Na_{0,5})NbO_3$. 2012
ISBN 978-3-86644-867-4

Band 5 Holger Schwaab
Nichtlineare Modellierung von Ferroelektrika unter Berücksichtigung der elektrischen Leitfähigkeit. 2012
ISBN 978-3-86644-869-8

Band 6 Christian Dethloff
Modeling of Helium Bubble Nucleation and Growth in Neutron Irradiated RAFM Steels. 2012
ISBN 978-3-86644-901-5

Band 7 Jens Reiser
Duktilisierung von Wolfram. Synthese, Analyse und Charakterisierung von Wolframlaminaten aus Wolframfolie. 2012
ISBN 978-3-86644-902-2

Band 8 Andreas Sedlmayr
Experimental Investigations of Deformation Pathways in Nanowires. 2012
ISBN 978-3-86644-905-3

Band 9 Matthias Friedrich Funk
Microstructural stability of nanostructured fcc metals during cyclic deformation and fatigue. 2012
ISBN 978-3-86644-918-3

Band 10 Maximilian Schwenk
Entwicklung und Validierung eines numerischen Simulationsmodells zur Beschreibung der induktiven Ein- und Zweifrequenzrandschichthärtung am Beispiel von vergütetem 42CrMo4. 2012
ISBN 978-3-86644-929-9

Band 11 Matthias Merzkirch
Verformungs- und Schädigungsverhalten der verbundstranggepressten, federstahldrahtverstärkten Aluminiumlegierung EN AW-6082. 2012
ISBN 978-3-86644-933-6

Band 12 Thilo Hammers
Wärmebehandlung und Recken von verbundstranggepressten Luftfahrtprofilen. 2013
ISBN 978-3-86644-947-3

Band 13 Jochen Lohmiller
Investigation of deformation mechanisms in nanocrystalline metals and alloys by in situ synchrotron X-ray diffraction. 2013
ISBN 978-3-86644-962-6

Band 14 Simone Schreijäg
Microstructure and Mechanical Behavior of Deep Drawing DC04 Steel at Different Length Scales. 2013
ISBN 978-3-86644-967-1

Band 15 Zhiming Chen
Modelling the plastic deformation of iron. 2013
ISBN 978-3-86644-968-8

Band 16 Abdullah Fatih Çetinel
Oberflächendefektausheilung und Festigkeitssteigerung von niederdruckspritzgegossenen Mikrobiegebalken aus Zirkoniumdioxid. 2013
ISBN 978-3-86644-976-3

Band 17 Thomas Weber
Entwicklung und Optimierung von gradierten Wolfram/EUROFER97-Verbindungen für Divertorkomponenten. 2013
ISBN 978-3-86644-993-0

Band 18 Melanie Senn
Optimale Prozessführung mit merkmalsbasierter Zustandsverfolgung. 2013
ISBN 978-3-7315-0004-9

Band 19 Christian Mennerich
Phase-field modeling of multi-domain evolution in ferromagnetic shape memory alloys and of polycrystalline thin film growth. 2013
ISBN 978-3-7315-0009-4

Band 20 Spyridon Korres
On-Line Topographic Measurements of Lubricated Metallic Sliding Surfaces. 2013
ISBN 978-3-7315-0017-9

Band 21 Abhik Narayan Choudhury
Quantitative phase-field model for phase transformations in multi-component alloys. 2013
ISBN 978-3-7315-0020-9

Band 22 Oliver Ulrich
Isothermes und thermisch-mechanisches Ermüdungsverhalten von Verbundwerkstoffen mit Durchdringungsgefüge (Preform-MMCs). 2013
ISBN 978-3-7315-0024-7

Band 23 Sofie Burger
High Cycle Fatigue of Al and Cu Thin Films by a Novel High-Throughput Method. 2013
ISBN 978-3-7315-0025-4

Band 24 Michael Teutsch
Entwicklung von elektrochemisch abgeschiedenem LIGA-Ni-Al für Hochtemperatur-MEMS-Anwendungen. 2013
ISBN 978-3-7315-0026-1

Band 25 Wolfgang Rheinheimer
Zur Grenzflächenanisotropie von $SrTiO_3$. 2013
ISBN 978-3-7315-0027-8

Band 26 Ying Chen
Deformation Behavior of Thin Metallic Wires under Tensile and Torsional Loadings. 2013
ISBN 978-3-7315-0049-0

Band 27 Sascha Haller
Gestaltfindung: Untersuchungen zur Kraftkegelmethode. 2013
ISBN 978-3-7315-0050-6

Band 28 Stefan Dietrich
Mechanisches Verhalten von GFK-PUR-Sandwichstrukturen unter quasistatischer und dynamischer Beanspruchung. 2016
ISBN 978-3-7315-0074-2

Band 29 Gunnar Picht
Einfluss der Korngröße auf ferroelektrische Eigenschaften dotierter $Pb(Zr_{1-x}Ti_x)O_3$ Materialien. 2013
ISBN 978-3-7315-0106-0

Band 30 Esther Held
Eigenspannungsanalyse an Schichtverbunden mittels inkrementeller Bohrlochmethode. 2013
ISBN 978-3-7315-0127-5

Band 31 Pei He
On the structure-property correlation and the evolution of Nanofeatures in 12-13.5% Cr oxide dispersion strengthened ferritic steels. 2014
ISBN 978-3-7315-0141-1

Band 32 Jan Hoffmann
Ferritische ODS-Stähle – Herstellung, Umformung und Strukturanalyse. 2014
ISBN 978-3-7315-0157-2

Band 33 Wiebke Sittel
Entwicklung und Optimierung des Diffusionsschweißens von ODS Legierungen. 2014
ISBN 978-3-7315-0182-4

Band 34 Osama Khalil
Isothermes Kurzzeitermüdungsverhalten der hoch-warmfesten Aluminium-Knetlegierung 2618A (AlCu2Mg1,5Ni). 2014
ISBN 978-3-7315-0208-1

Band 35 Magalie Huttin
Phase-field modeling of the influence of mechanical stresses on charging and discharging processes in lithium ion batteries. 2014
ISBN 978-3-7315-0213-5

Band 36 Christoph Hage
Grundlegende Aspekte des 2K-Metallpulverspritzgießens. 2014
ISBN 978-3-7315-0217-3

Band 37 Bartłomiej Albiński
Instrumentierte Eindringprüfung bei Hochtemperatur für die Charakterisierung bestrahlter Materialien. 2014
ISBN 978-3-7315-0221-0

Band 38 Tim Feser
Untersuchungen zum Einlaufverhalten binärer alpha-Messinglegierungen unter Ölschmierung in Abhängigkeit des Zinkgehaltes. 2014
ISBN 978-3-7315-0224-1

Band 39 Jörg Ettrich
Fluid Flow and Heat Transfer in Cellular Solids. 2014
ISBN 978-3-7315-0241-8

Band 40 Melanie Syha
Microstructure evolution in strontium titanate Investigated by means of grain growth simulations and x-ray diffraction contrast tomography experiments. 2014
ISBN 978-3-7315-0242-5

Band 41 Thomas Haas
Mechanische Zuverlässigkeit von gedruckten und gasförmig abgeschiedenen Schichten auf flexiblem Substrat. 2014
ISBN 978-3-7315-0250-0

Band 42 Aron Kneer
Numerische Untersuchung des Wärmeübertragungsverhaltens in unterschiedlichen porösen Medien. 2014
ISBN 978-3-7315-0252-4

Band 43 Manuel Feuchter
Investigations on Joule heating applications by multiphysical continuum simulations in nanoscale systems. 2014
ISBN 978-3-7315-0261-6

Band 44 Alexander Vondrous
Grain growth behavior and efficient large scale simulations of recrystallization with the phase-field method. 2014
ISBN 978-3-7315-0280-7

Band 45 Tobias Kennerknecht
Fatigue of Micro Molded Materials – Aluminum Bronze and Yttria Stabilized Zirconia. 2014
ISBN 978-3-7315-0293-7

Band 46 Christopher Scherr
Elektrochemisches Verhalten von Lithium-Schwefel-Zellen mit unterschiedlicher Kathodenstruktur. 2015
ISBN 978-3-7315-0296-8

Band 47 Konstantin Frölich
Der Decal-Prozess zur Herstellung katalysatorbeschichteter Membranen für PEM-Brennstoffzellen. 2015
ISBN 978-3-7315-0334-7

Band 48 Benedikt Haspel
Werkstoffanalytische Betrachtung der Eigenschaften von mittels neuartiger RTM-Fertigungsprozesse hergestellten glasfaserverstärkten Polymerverbunden. 2015
ISBN 978-3-7315-0337-8

Band 49 Marco Berghoff
Skalenübergreifende Modellierung und Optimierung vom atomistischen kristallinen Phasenfeldmodell bis zur mesoskopischen Phasenfeldmethode. 2015
ISBN 978-3-7315-0416-0

Band 50 Michael Selzer
Mechanische und Strömungsmechanische Topologieoptimierung mit der Phasenfeldmethode. 2016
ISBN 978-3-7315-0431-3

Band 51 Michael Mahler
Entwicklung einer Auswertemethode für bruchmechanische Versuche an kleinen Proben auf der Basis eines Kohäsivzonenmodells. 2016
ISBN 978-3-7315-0441-2

Band 52 Christoph Bohnert
Numerische Untersuchung des Verformungs- und Bruchverhaltens von einkristallinem Wolfram auf mikroskopischer Ebene. 2016
ISBN 978-3-7315-0444-3

Band 53 Stefan Guth
Schädigung und Lebensdauer von Nickelbasislegierungen unter thermisch-mechanischer Ermüdungsbeanspruchung bei verschiedenen Phasenlagen. 2016
ISBN 978-3-7315-0445-0

Band 54 Markus Klinsmann
The Effects of Internal Stress and Lithium Transport on Fracture in Storage Materials in Lithium-Ion Batteries. 2016
ISBN 978-3-7315-0455-9

Band 55 Thomas Straub
Experimental Investigation of Crack Initiation in Face-Centered Cubic Materials in the High and Very High Cycle Fatigue Regime. 2016
ISBN 978-3-7315-0471-9

Band 56 Maren Lepple
Kupfer- und Eisenoxide als Konversions-Elektrodenmaterialien für Lithium-Ionen-Batterien: Thermodynamische und Elektrochemische Untersuchungen. 2016
ISBN 978-3-7315-0482-5

Band 57 Stefan Andreas Slaby
Charakterisierung und Bewertung der Zug- und Ermüdungseigenschaften von Mikrobauteilen aus 17-4PH Edelstahl. Ein Vergleich von mikropulverspritzgegossenem und konventionell hergestelltem Material. 2017
ISBN 978-3-7315-0484-9

Band 58 Kumar Ankit
Phase-field modeling of microstructural pattern formation in alloys and geological veins. 2016
ISBN 978-3-7315-0491-7

Band 59 Kuo Zhang
Characterization and Modeling of the Ratcheting Behavior of the Ferritic-Martensitic Steel P91. 2017
ISBN 978-3-7315-0503-7

Band 60 Jens Bauer
Mechanische Metamaterialien – Hohe Festigkeit und niedrige Dichte durch Mikroarchitektur und Materialdesign. 2017
ISBN 978-3-7315-0506-8

Band 61 Fabian Lemke
Untersuchung des Sinterverhaltens von $SrTiO_3$ unter Berücksichtigung der Defektchemie. 2016
ISBN 978-3-7315-0510-5

Band 62 Johannes Kümmel
Detaillierte Analyse der Aufbauschneidenbildung bei der Trockenzerspanung von Stahl C45E mit Berücksichtigung des Werkzeugverschleißes. 2016
ISBN 978-3-7315-0518-1

Band 63 László Hagymási
Modellierung der Stoffübertragung beim Niederdruckcarbonitrieren mit Ammoniak und Acetylen. 2016
ISBN 978-3-7315-0568-6

Band 64 Reza Eslami
A novel micro-mechanical model for prediction of multiaxial high cycle fatigue at small scales. 2017
ISBN 978-3-7315-0583-9

Band 65 Sebastian Schulz
Phase-field simulations of multi-component solidification and coarsening based on thermodynamic datasets. 2017
ISBN 978-3-7315-0618-8